Joseph B. Murdock

Notes on Electricity and Magnetism

Designed as a companion to Silvanus P. Thompson's Elementary Lessons

Joseph B. Murdock

Notes on Electricity and Magnetism
Designed as a companion to Silvanus P. Thompson's Elementary Lessons

ISBN/EAN: 9783337275693

Printed in Europe, USA, Canada, Australia, Japan

Cover: Foto ©berggeist007 / pixelio.de

More available books at **www.hansebooks.com**

NOTES

ON

ELECTRICITY AND MAGNETISM.

DESIGNED AS A COMPANION TO SILVANUS P. THOMPSON'S
ELEMENTARY LESSONS.

BY

J. B. MURDOCK,

LIEUTENANT U. S. NAVY.

New York:
MACMILLAN & CO.
1891.

PREFACE.

THE design of this small volume of notes is to supplement the instruction given in Prof. Thompson's admirable Elementary Lessons by such explanations and additional instruction as an experience with two classes of cadets in this institution has shown to be necessary. In general, notes have been made on separate paragraphs in the lessons, but it was thought best to treat many subjects independently in order to present them more connectedly, and it is hoped that the notes may thus be of some service by themselves. The endeavor has been made to trace the theory of the dynamo machine and of electric motors from the primary laws of electro-magnetic induction, and descriptions of several of the more important dynamo machines have been added, chiefly to illustrate the various applications of the general principle underlying all. As Prof. Thompson's treatise has come to be largely used in colleges and high schools, demonstrations by the aid of calculus, most of which have been used in the course in this institution, have been given to replace the geometrical proofs of the lessons if desired.

J. B. MURDOCK.

U. S. NAVAL ACADEMY,
Annapolis, Md., *July* 10, 1883.

NOTE.—Reference have been made in the text to the figures in the Elementary Lessons, as well as to those in this volume. The latter range from 1 to 38, and higher numbers are to be understood as referring to those in the Elementary Lessons.

INDEX TO NOTES.

III. Theory of Magnetic Potential.

IV. Measurements and Formulas.

NOTES

ON

ELECTRICITY AND MAGNETISM.

I. GALVANOMETERS.

(Thompson's Electricity, pages 163–171.)

1. Tangent Galvanometer (§ 199).

To derive the formula in § 200. Let the magnet be deflected by the current to an angle θ with the meridian and be in equilibrium. The sum of the moments of the forces acting on it, taken around its axis, is then zero. The moment tending to bring it back into the meridian is the force mH into the arm BC, or $mlH \sin \theta$, m being the strength of magnet pole and l the distance between the poles. The moment of the deflecting couple is $fm \times DE$, or $fml \cos \theta$, f being the deflecting force of the current. If now the magnet be so small that in all positions its poles may be considered to be at the centre of the coil $f =$

Fig. 1.

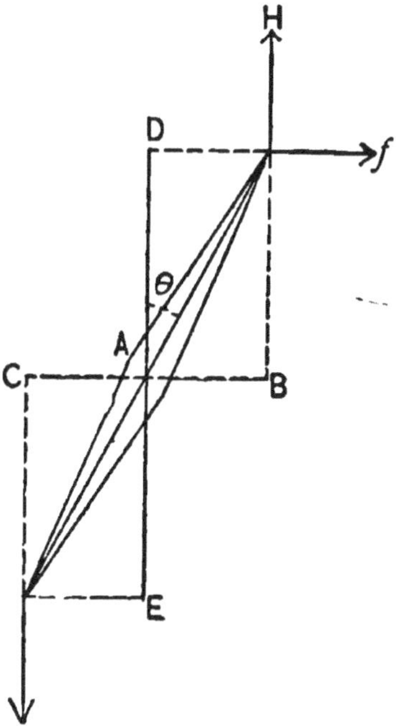

$\dfrac{2\pi C}{r}$ (§ 195). If the coil has n turns, each exerting this force on a pole at the centre, $f = \dfrac{2\pi n C}{r}$. Substituting this value of f and equating the moments of the forces around A

$$mlH \sin \theta = \frac{2\pi n C}{r} \, ml \cos \theta$$

$$\therefore C = \frac{r}{2\pi n} H \tan \theta.$$

This formula shows not only that currents vary as the tangent of the angle of deflection, but gives the value of the currents in the absolute units of current defined in § 196. The practical unit of current, the ampère, being only $\frac{1}{10}$ of the absolute unit, the result must be multiplied by ten to obtain the current in ampères from any observed deflection. The fraction $\dfrac{r}{2\pi n}$ depending on the construction of the instrument, is called the " reduction factor " of the galvanometer. It is generally furnished by the maker, but may be readily determined by electrolysis. If an electrolytic cell and a galvanometer are in the same circuit, as the current is the same throughout, the value of C in ampères, as given by the tangent galvanometer, may be equated with that from the equation on page 177. Thus

$$\frac{10w}{zt} = \frac{r}{2\pi n} H \tan \theta$$

$$\therefore \text{Reduction factor} = \frac{10w}{Hzt \tan \theta}.$$

The demonstration given above shows that the deflection is independent of the strength of pole, and it is not, therefore, necessary to magnetize the needle strongly. As the demonstration involves a value of f which is true only when the pole is at the centre of the coil, the magnet must be so small, relatively to the radius, that neither of its poles should in any position depart widely from the centre. As the lines of force due to a current in a coil pass through its plane perpendicularly, as shown in Fig. 86, the dimensions of the needle and the accuracy of the instrument may be increased by using two coils with the needle midway on their common axis. The lines of force of the two coils then act together so that they are sensibly parallel in the region in which the needle moves, many of them passing through both coils perpendicularly.

2. Sine Galvanometer (§ 201).

It is possible to construct a sine galvanometer to measure currents in absolute units, but the ordinary form of the instrument is not intended for absolute but for relative values only. The coil is placed parallel to the needle before the observation, and when a deflection has been produced by the passage of a current, the coil is rotated and the attempt made to bring it once more parallel to the needle. Every movement of the coil produces a further displacement of the needle and the two are brought into the same vertical plane only by careful adjustment. When the coil and needle are parallel, the moment of the earth's directive force is, as before, $mlH \sin \theta$, but the deflecting influence of the current, acting always at right angles, is fml

Equating $\qquad f = H \sin \theta.$

In the same galvanometer, f is always a function of C, and hence the current varies as the sine of the angle of deflection. Thus knowing the deflection that a current of known strength produces

$$C : C' : : \sin \theta : \sin \theta'.$$

No current can produce a deflection of more than $90°$

in a tangent galvanometer or in one in which the coil is fixed. In a sine galvanometer, however, as the coil is always kept parallel to the needle it exerts the same deflecting force in all positions. If now the current is of such a strength that equilibrium is attained at 90° from the original position of the needle, a stronger current would deflect it still farther, and it would be impossible to obtain equilibrium.

3. Mirror Galvanometer (§ 202).

The supposition is generally made that with the mirror galvanometer the currents vary directly as the scale readings, but this is true only within limits. The needle being small and the coil large, the current is proportional to the tangent of the deflection, but as the deflection is read by the movement of a spot of light on a tangent scale, the current would be proportional to the reading, if it did not follow from the laws of reflection that the spot of light moved over twice the angle that the mirror moved. Calling the observed deflections d and d', the true ratio is

$$C : C' :: \tan \frac{d}{2} . \tan \frac{d'}{2}.$$

In assuming that the currents are proportional to the readings, the supposition is ·

$$C : C' :: r : r' :: \tan d : \tan d'.$$

If $\tan d : \tan d' :: \tan \frac{d}{2} : \tan \frac{d'}{2}$, as it is sensibly with small values of d and d', the deflections may be taken to be proportional to the strengths of the currents. In the mirror galvanometer, therefore, the currents are proportional to the readings if the deflections are small.

4. Differential Galvanometer (§ 203).

The differential galvanometer is used not to measure

currents, but to indicate that two are either equal or unequal. As the greater the resistance a current has to
flow through, the smaller the current is, by changing the
resistances in the circuit of either of the two coils of the
galvanometer the currents may be made equal. It af-
fords, therefore, an easy method of comparing resistances.

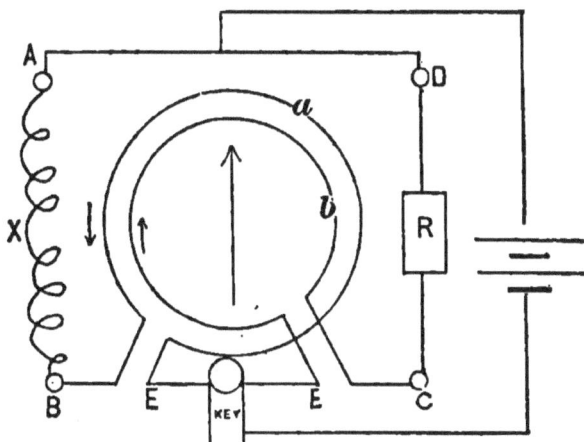

Fig. 2.

In the figure it is seen that if the key is depressed to make
contact with EE, the currents pass around the needle in
opposite directions. In the figure one current is farther
from the needle than the other, but great care is taken in
making the instrument to have the coils similarly placed
and of equal resistance, so that equal currents will flow
through them, producing also equal but opposite effects
on the magnet. If there is no deflection, the currents are
equal, or $C = C'$;

$$\text{but } C = \frac{E}{R + a} \quad \text{and} \quad C' = \frac{E}{x + b}.$$

But since $a = b$, $x = R$, if there is no deflection of the
needle when the key is pressed. By having R adjustable, x may be determined.

5. Ballistic Galvanometer (§ 204).

When a current is of very short duration, it may be supposed to exert an impulsive force on a galvanometer needle, especially if the latter is heavy, as it would then, by virtue of its inertia, fail to move until all the varying impulses of the transient current had been given it. The effect is then, practically, that produced by a momentary impulse, and the needle will move with a velocity proportional to the quantity passing, the force exerted in any case varying as the current and the time the current lasts, or as the quantity if the time is small, and will vibrate through a certain arc, coming to rest, and then under the directive action of the earth's magnetism will make a return oscillation acquiring the same velocity as that originally given by the current. In this oscillation the effective force tending to bring it to rest varies as the sine of the angle of deflection, as it does also in the case of a simple pendulum, and the relations deduced from the latter may therefore be applied without perceptible error to the vibrating magnet.

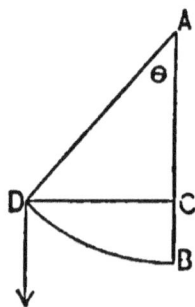

The velocity acquired by the pendulum in falling through the arc $D\,B$ is equal to that it would acquire in falling from C to B; hence from the laws of falling bodies

$$v = \sqrt{2gBC}$$

But $BC = AD\,(1-\cos\theta)=2\,AD\sin^2\tfrac{1}{2}\theta$ calling $AD,\ l$

$$v = \sqrt{4\,lg\sin^2\tfrac{1}{2}\theta}$$

$$v = 2\sin\tfrac{1}{2}\theta\,\sqrt{gl}\cdot$$

Fig. 3.

But as the velocity is proportional in the galvanometer to

the quantity passing, the quantity varies as the sine of half the angle of deflection.   If the deflection is noted when a known quantity is discharged through a ballistic galvanometer, any other quantity may be determined by taking the ratio of the sines of half the respective deflections.

II. THEORY OF POTENTIAL.

(Thompson's Electricity, pages 190–208.)

6. Potential (§ 237).

An explanation of the term potential must precede any definition. To illustrate, suppose a weight of a pound be moved by the hand. In order to lift it to a higher level the muscles have to be called on to do work. If the pound is lifted ten feet, it is only by the expenditure of ten foot pounds of work. If after being lifted, it is placed on a shelf, the work done on it is evidently in the form of potential energy, and may be recovered if the weight is allowed to fall, when it will do ten foot pounds of work. Strictly speaking, the weight on the shelf possesses energy by virtue of the work done on it, but if this work had not been done visibly, the weight might be said to possess energy by virtue of its mass and the *potential* or *height* to which it had been raised. A stone on the top of a precipice is thus said to have energy by virtue of its potential, disregarding the question of how much work is required to lift it to its position. Let a weight of two pounds be now lifted to the same shelf, at a height or potential of ten feet. If allowed to fall it would do twenty foot pounds of work, although falling from the same height or potential as did the other weight which performed only ten foot pounds in its descent. The work has been doubled, although the potential is the same. The work done, either on the weight in lifting it to the shelf, or by the weight in falling, is evidently the product of the weight into the potential.

If a common bar magnet be taken in the hand and moved near a powerful fixed magnet, work will be done. If like poles are near each other, a force of repulsion is exerted, and the muscles are called upon to do work in bringing the magnet nearer to the fixed magnet, moving it against this force. If the bar magnet be now suspended so as to be unable to turn end for end, it will when released by the hand fly away from the fixed magnet under the influence of the force of repulsion, doing work in its movement. The work done against the magnetic forces in bringing the bar magnet nearer the fixed magnet becomes potential energy, and is available as kinetic whenever the restraining force of the hand is removed. Here, as in the case of the weight, the potential energy is derived from the work previously done in moving the magnet, but it is simpler to say that the magnet possesses energy by virtue of the potential to which it is raised. If another magnet of double the strength be moved, twice as much work will have to be expended on it in bringing it to the position occupied by the first, and it will have twice the potential energy. As it has been brought to the same potential the double work is due to its being of twice the strength. In this case, therefore, work is the product of strength of pole and magnetic potential.

Suppose a unit quantity of positive electricity be moved near a larger quantity also positive. A force of repulsion exists between them, and if they be brought nearer together, the unit quantity will if released fly away from the larger quantity, doing work. It evidently possesses potential energy, or does work by virtue of its potential. If a charge of two units be brought up twice the work will be done, and there will be twice the potential energy. Here work is evidently the product of quantity of electricity and electrostatic potential.

7. Difference between Work and Potential.

As in each of the three cases examined the work done or the potential energy possessed is the product of potential and some other factor, if that factor be known the potential may be obtained by measuring the work. Potential is therefore *measured* by work, but is not work. In several places in " Thompson's Electricity " the statement is made that " potential is the work." The difference between them may appear more clearly from a recapitulation of the relations already traced.

Gravitation Potential $= \dfrac{\textit{Work done in lifting weight}}{\textit{Weight lifted.}}$

Magnetic Potential $=$

$\dfrac{\textit{Work done in moving magnet pole}}{\textit{Strength of pole.}}$

Electrostatic Potential $=$

$\dfrac{\textit{Work done in moving quantity of electricity}}{\textit{Quantity moved.}}$

Having thus traced the general analogies, in further consideration electrostatic potential alone need be considered. From the last equation it is seen that if unit quantity be moved, the potential is numerically equal to the work done in moving it. As the work done in moving from zero potential is the measure of the potential energy acquired, *the electrostatic potential at a point equals the potential energy possessed by unit quantity of positive electricity at that point, and is measured by the work that must be spent in bringing unit quantity of positive electricity up to the point from an infinite distance.* The infinite distance enters the definition from the fact that there the potential is zero. This is evident from the fact that potential is measured by work ; work is the product of force into distance, over which it acts, and force $= \dfrac{qq'}{r^2}$. If r is infinite, the force is zero, and no work is done in moving unit quantity.

8. Positive and Negative Work.

As in any case in which a positive unit is repelled, a negative unit is attracted, the work which in the first case is necessary to move the unit against the force of repulsion would in the second be done in preventing the movement of the unit under the forces of attraction.   The two cases are evidently diametrically opposite, and the work done is therefore considered as positive when it is done *on* the the unit, negative when done *by* the unit moving freely. It becomes necessary, therefore, to specify the positive unit in the definition, that the nature of the work and consequently the sign of the potential may be known.

9. Positive Electricity always flows from a High to a Low Potential.

Potential energy always tends to run down to a minimum.   A weight acted on by gravity will fall to the earth if not prevented ; a magnet pole placed near another similar pole possesses potential energy and tends to move away into a position in which its potential energy is less ; a unit of electricity placed near a similar quantity is likewise repelled and moves so as to decrease its potential.   In any electrified region the relative potential is therefore indicated by the direction in which a unit of positive electricity tends to move, and the distribution of potential may be examined by conceiving a positive unit to be moved throughout the neighborhood of the electrified bodies, noting whether it is necessary to do work to move it or to restrain its movement. Let this positive unit be approached to a quantity of positive electricity. Work must be done to move it, and if left free it will fly away, moving to decrease its potential. It has evidently been moved into a region of higher potential. If approached to a negative unit, work must be done to restrain its movement. If free to move it will move toward the negative unit, and is moving into a region of lower potential. Generally speaking, a body positively electrified is at a positive potential, and one negatively electrified at a negative, but there are many exceptions.

Suppose a cylinder B to be unelectrified and to be connected with the earth by a wire. There is no flow of electricity in the wire, and B is therefore at the same potential as the earth, as electricity tends to move toward a lower potential,

Fig. 4.

and the fact of there being no movement shows there is no difference of potential. Let a positively electrified ball A be approached, and B becomes electrified by induction as in the figure. If again connected with the earth, a flow of positive electricity takes place from the cylinder to the earth, showing that the potential of B had been *raised* by the approach of A. Before being connected with the earth the second time it was therefore at a positive potential but was negatively electrified at one end. When in contact with the earth, with A as in the figure, it is negatively electrified, but at zero potential.

10. Units of Potential and Work.

Potential is measured by work, and the units of potential are numerically equal to the units of work. Work is defined as force acting through distance, and the C. G. S. unit of work called the *erg* is the work done in opposing the force of one dyne through the distance of one centimetre. It may be defined after the analogy of foot-pounds as a dyne-centimetre. If, therefore, the work necessary to move a unit quantity of electricity, or the work a unit quantity does in moving, is measured in *ergs*, it numerically equals the difference of potential through which the unit moves.

11. Electrostatic Potential (§ 238).

We can now derive the general formula for electrostatic potential. From the definitions of potential and work—

$$\text{Work} = \int_r^\infty f dr, \text{ where } r \text{ is distance ;}$$

$$\text{but } f = \frac{qq'}{r^2} \quad \text{and Work} = \int_r^\infty \frac{qq'}{r^2}\, dr.$$

If q' is unity, work measures potential, hence, denoting potential by V,

$$V = \int_r^\infty \frac{q dr}{r^2} = \frac{q}{r} \cdot$$

The potential at any point due to a quantity q is, therefore, numerically equal to the quantity divided by the distance in centimetres. If other quantities q', q'', etc., were near, the potential due to them would be $\frac{q'}{r'}$, $\frac{q''}{r''}$, etc. The potential due to the whole system is then

$$V = \frac{q}{r} + \frac{q'}{r'} + \frac{q''}{r''} = \Sigma \frac{q}{r}.$$

If either q, q' or q'' is negative, $\dfrac{q}{r}$, $\dfrac{q'}{r}$ or $\dfrac{q''}{r}$ will be negative, and must be given its proper sign in the summation.

12. Zero Potential (§ 239).

Although as shown, the theoretical zero potential exists at an infinite distance, the potential of the earth at the place is the practical zero. All electrical manifestations are dependent on a difference of potential, and the absolute potential is never needed.

13. Difference of Potentials (§ 240).

Potential being measured by work done, the difference of potential between two points is numerically equal to the number of ergs required to move a positive unit from one point to the other. It is immaterial what path be followed, as if all the work done, both positive and negative, be summed up, it will be equal to that done in moving in a direct line between the points. Two quantities of electricity at different potentials may be compared to two ponds of water at different levels. If the ponds are connected by a pipe, the water in the upper will by virtue of its height possess potential energy and will run down into the lower. If no current of water flowed in the pipe it would indicate that the ponds were at the same level. If no other means of measuring the difference of level were available, it could be done by measuring in foot-pounds the work done by one pound of water in flowing through the pipe. Similarly electricity flows from a body electrified to a high potential to one at a lower connected with it, and the difference of potential is measured by the work in ergs done by unit quantity in flowing from one to the other.

14. Electric Force (§ 241).

When f is force exerted on unit quantity

$$V = \int f dr, \text{ or } dV = f dr \quad \therefore f = \frac{dV}{dr}.$$

But $\frac{dV}{dr}$ is the rate of change of potential, hence the average electric force between two points at different potentials is measured by the rate of change of potential per centimetre.

15. Law of Inverse Squares (§ 245, Fig. 98).

Coulomb's observations, without proving exactly that the law of inverse squares applied to electric force, so nearly proved it as to lead one to think that more careful experimentation, were such possible, would demonstrate the exactness of the law.   Assume, therefore, the law and trace the results.

Let ρ be the electric density, or the amount of electricity per square centimetre of surface.   On a sphere removed from other conductors the density is uniform, and the quantity on two surfaces varies as the area of the surfaces. The quantity on the surface AB (Fig. 98, " Thompson ") is $\rho \times$ Area $AB = \rho A$.

The quantity on CD is $\rho \times$ Area $CD = \rho C$.

Assuming that electric force varies inversely as the square of the distance, the force exerted on a unit of electricity at the point P inside the sphere by the quantity on AB is

$$f = \frac{\rho A}{\overline{BP}^2}.$$

The force exerted on the same unit by the quantity on CD is

$$f' = \frac{\rho C}{\overline{CP}^2}.$$

Since the tangents drawn to the sphere are equally inclined to BC

$$A : C :: \overline{BP}^2 : \overline{CP}^2, \quad \text{or } A = \frac{C \times \overline{BP}^2}{\overline{CP}^2},$$

substituting above $f = f'$.

If the sphere be cut up into small cones $\Sigma f = \Sigma f'$; or in other words, if electric force varies inversely as the square of the distance, there should be no resultant force on the inside of a closed conductor. If it follows any other law f cannot equal f'. The most careful experiments fail to detect any force existing, and in corroborating the result of the above demonstration, confirm the hypothesis made that electric force follows the law of inverse squares.

16. Capacity ($\frac{3}{3}$ 246).

The capacity of a conductor is by the definition given a fixed quantity, while "the amount of electricity the conductor can hold," the definition as generally given by beginners, is variable, depending on the potential as well as on the capacity. An illustration may make the distinction clearer. If a jar has a volume of one litre, its *capacity* is a litre, and it will hold a litre of air at atmospheric pressure. If the pressure be doubled, however, the quantity of air in the jar is also doubled, although the capacity is the same. "The amount" the jar "will hold" is evidently determined by the *capacity* and *pressure*. If the pressure be unity, that of one atmosphere, the capacity is then the amount the jar actually holds, but quantity and capacity are under other conditious different. With reference to electricity, the capacity is similarly the charge the conductor will hold at unit potential, or the charge which will raise, the potential to unity, and the actual charge in a conductor is the product of the capacity and potential.

17. Unit of Capacity (§ 247).

In an electrified sphere, as the surface is an equipotential surface, the charge may be considered as concentrated at the centre and the potential at the surface is $\frac{q}{r}$. As the capacity is equal to the quantity divided by the potential

$$\text{Capacity} = \frac{q}{\frac{q}{r}} = r.$$

A sphere of one centimetre radius is, therefore, of unit capacity.

18. Electric Force exerted by a Charged Plate (§ 252).

Let a be the radius of the plate and r the radius of the ring $x, x,' x.''$ ρ is the density, or the charge per unit of area.

The quantity on any small circular element is $\rho (2\pi r dr)$. The force exerted by this quantity on a unit at O is

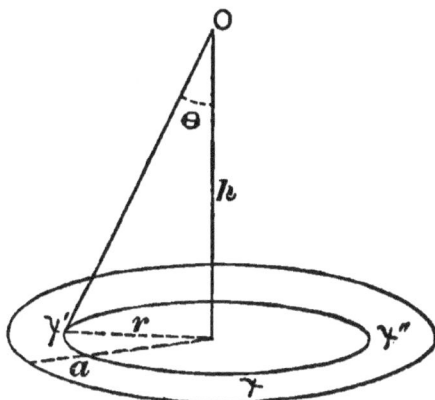

Fig. 5

$$\frac{\rho(2\pi r dr)}{h^2 + r^2};$$

and the force acting normal to the plate is

2

$$\rho\left(\frac{2\pi r dr}{h^2 + r^2}\cdot\frac{h}{\sqrt{h^2 + r^2}}\right),\quad\text{since } \cos\theta = \frac{h}{\sqrt{h^2 + r^2}}.$$

The total force exerted by the plate in a direction normal to its surface is, therefore

$$\rho\int_0^a\frac{2\pi r dr}{(h^2 + r^2)}\cdot\frac{h}{(h^2 + r^2)^{\frac{1}{2}}} = \pi\rho h\int_0^a\frac{2\,r dr}{(h^2 + r^2)^{\frac{3}{2}}}$$

Integrating,

$$= -2\pi\rho h\,(h^2 + r^2)^{\frac{1}{2}}\Big]_0^a$$

$$= \frac{2\pi\rho h}{h} - 2\pi\rho\,,\frac{h}{(h^2 + a^2)^{\frac{1}{2}}}$$

$$= 2\pi\rho\,(1 - \cos\theta').$$

If O is very near the plate, or if the plate is very large, $\theta' = 90°$, and the electric force of a charged plate on a unit of electricity very near it is $2\pi\rho$.

Care must be taken not to confuse this with the force exerted by a sphere as deduced in § 251. If a plate is charged with positive electricity, and a positive unit is placed very near it on each side, the force will be one of repulsion in each case, but if one unit is repelled upward the other tends to move downward, and if one force is $2\pi\rho$, the other must be $-2\pi\rho$. The force changes, therefore, by $4\pi\rho$ in passing through any charged surface.

19. Dimensions of Units (§ 258).

An important use of the dimensional equations is in the conversion of units based on one system of fundamental units of mass, length and time to others based on different fundamental units. The French use units based on the metrical system, and although the centimetre-gramme-second, or C. G. S. system, is now almost universally used in electrical work, there are many observations made and recorded in which other units are used. In electrical work the English have heretofore used the foot-grain-second system, and it is still used in some government observatories. It is a matter of the highest importance, therefore, that the method of converting values expressed in one system to corresponding values in another should be thoroughly understood. As the ratio of the centimetre to the foot is that of 1 to 30.48, it is evident that the ratio of the units of length in the C. G. S. and foot-grain-second systems is the same. The units of area in the two systems are the square centimetre and the square foot, and no one would think of saying that this ratio was the same as the preceding $\frac{1}{30.48}$, but rather $\frac{1^2}{(30.48)^2}$. Similarly the ratio between the units of volume is not $\frac{1}{30.48}$ but $\frac{1^3}{(30.48)^3}$. This is exactly the relation shown by the dimensions of area and volume on page 211, they being respectively L^2 and L^3. In simple cases like the above the change is easily made, but in others, where the dimensions of the unit are more complex and the unit itself an unfamiliar one, the dimensions must be used to calculate the ratio. As an illustration, let it be required to express in units of potential based on the foot, grain and second, the difference of electrostatic potential expressed by 2.7 C. G. S. units of potential.

The dimensions of electrostatic potential are $M^{\frac{1}{2}} L^{\frac{1}{2}} T^{-1}$,

then $\dfrac{\text{The } C. G. S. \text{ unit}}{\text{Foot-Grain-Sec. unit}} = \left(\dfrac{m}{M}\right)^{\frac{1}{2}} \left(\dfrac{l}{L}\right)^{\frac{1}{2}}$

$$= \left(\frac{15.43}{1} \times \frac{1}{30.48}\right)^{\frac{1}{2}} = .7116$$

∴ 1 *C. G. S.* unit = .7116 Foot-grain-second unit
2. 7 *C. G. S.* = 1.92 Foot-grain sec.

The ratios between the different units in the two systems are given in Note 40.

20. Attracted Disc Electrometers (§ 261, p. 215).

Let the difference of potential between two plates be V, and the distance apart be D. By Note 14 the average electric force between the plates is $\frac{V}{D}$. As proved in Note 18, the electric force changes by $4\pi\rho$ in passing through a surface, and being zero in the conductor is therefore $4\pi\rho$ just outside. Equating, $\rho = \frac{V}{4\pi D}$. The density on each plate being ρ, the attraction exerted by the lower plate on a unit of electricity on the upper one is $2\pi\rho$ when the plates are near each other. The upper plate contains, however, $S\rho$ units and the total attraction is $2\pi\rho \times S\rho = 2\pi S\rho^2$,

$$\therefore F = 2\pi S\rho^2 = 2\pi S \frac{V^2}{16\pi^2 D^2};$$

$$\therefore V = D \sqrt{\frac{8\pi F}{S}}.$$

21. Absolute Electrometer (p. 216).

Sir William Thompson's absolute electrometer, so named from giving the potential in absolute units, is an attracted disc electrometer.

The disc C (see Figure 100, "Thompson") is held in place by springs, instead of a counterpoise as shown, and is in metallic connection with B. When no part of the apparatus is electrified, small weights are placed on C, to bring it into a standard position such that a small hair attached to it is seen midway between two dots, as shown in the figure. The weights are then removed and B and C connected to one of the bodies whose difference of potential is required and A to the other. The electric force of attraction between the two plates will act to lower C, but as the accuracy of the instrument depends on its being in the plane of B, the plate A is moved up or down until the force of attraction is such as to bring the movable plate into this standard position, which is known by seeing the hair again midway between the dots. It is now under the attraction of the electrical forces, in exactly the same position as when acted upon by the weights, and the two forces are therefore equal. Substituting, therefore, for F in the formula of Note 20, 981 times the weight in grammes required to bring C into the standard position, and for D the distance in centimetres between A and C, all quantities in the equation are known and the difference of potential may be calculated.

Another method is more common, dispensing with the use of weights. If the difference of potential between two bodies P and P' is required, one of the bodies is connected to A, and B and C are then electrified to a high potential. The plate A is then moved up or down until the plate C comes into the standard position, the hair showing midway between the dots, and the distance D of A from C is noted. Then

$$\text{Potential of } B - \text{Potential of } P = D \sqrt{\frac{8\pi F}{S}}.$$

The plate A is next disconnected from the first body and connected with the second. As the difference of potential between A and C has now been changed, the force acting between the plates is different and C is no longer in the standard position. It is brought there by raising or lowering A, and when adjusted the distance D' between the plates is noted :

Then Potential of $B-$ Potential of $P' = D'\sqrt{\dfrac{8\pi F}{S.}}$

Subtracting this from the former,
Difference of Potential between P and $P' = (D-D') \times$ constant of instrument.

It is of course necessary that the potential of B should be the same in both cases. This is verified by a separate attracted disc, which is in each case electrified until its attraction for another disc at a fixed distance brings it into a standard position. The absolute potential of B is immaterial, the only requirement being that it should be the same in each case. If the absolute potential of P is wished, connect A first to P and then to earth.

III. THEORY OF MAGNETIC POTENTIAL.

(Thompson's Electricity, pages 265–278.)

22. Magnetic Field.

Any region throughout which forces act is called a "field," but the term is more frequently used in connection with magnetic than with other forces. A *magnetic field* is, therefore, a region in which magnetic effects are produced. Any movement of a magnet pole can take place only in a magnetic field, and the term is of use, as it disregards all ideas of how the field is caused, and considers only the forces and the direction in which they act. If at any point a line is drawn indicating the direction of the force at that point, it is called a line of force. This direction is that shown by a magnet placed at the point, and may therefore be easily determined by experiment ; but as any representation of a magnetic field must present the whole field at once, the determination of the position of the lines of force by this process would be tedious. The reasoning in § 126 leads to an easier though less accurate method, but one of great utility in enabling clear conceptions to be formed. If a magnet is covered by a sheet of paper and iron filings are sprinkled over the paper they will on being gently tapped arrange themselves in curves passing from pole to pole. From the definition of lines of force, these curves must be the lines of force in the plane of the paper, and the mind has only to conceive the space above and below the magnet to be similarly filled to gain a clear idea of the field. It is necessary, however, to know not only the direction of the force at any point, but also its *strength*, and a correct plotting of the field must furnish this. Maxwell has shown that if in any part of their course, the number of lines of force passing through unit area of a perpendicular plane is proportional to the strength of the force there, the number passing through unit area in any other part of the field is in the same proportion to the strength in that part. The closeness of the lines of force is therefore a measure of the strength of the forces of the field, or, as more commonly expressed, of the *intensity* of the field. By drawing the lines of force therefore in this way, the strength and direction of the forces in all parts of the field are indicated. As a south pole moves always in the opposite direction to that in which a north pole moves, it is necessary in order to establish the direction of the force to consider the nature of the pole acted upon. All investigations in magnetism are made by considering a north pole free to move, and *the positive direction of the lines of force is therefore that in which a free north pole moves.* This definition is of great importance in many of the demonstrations given later.

23. Mapping a Field by Lines of Force.

A magnetic field is of unit intensity when unit pole is acted upon by a force of one dyne. As by definition unit pole acts on an equal and similar pole at a distance of one centimetre with a force of one dyne, it follows that unit pole causes unit field at unit distance. As intensity of field is measured by the force acting on unit pole, unit field exists at a greater distance from a more powerful pole. It is, therefore, unnecessary to consider the question of distance from the pole producing the field, but simply bear in mind that the intensity of the field at any point is measured by the force in dynes acting on a unit pole at that point. If the pole be of a strength m, the force with which it is attracted or repelled is m times that experienced by unit pole, or

$$f = mH;$$

$$\text{Intensity of field} = \frac{\text{Force acting on pole}}{\text{Strength of pole}}.$$

The value of H, or the strength of field, is given numerically. Thus the horizontal force of the earth's magnetism at London being .18, a pole of unit strength is impelled to move in a horizontal plane by a force of .18 dynes. A pole of strength 100 would be acted on by a force of 18 dynes in a horizontal direction, or by a force of 47 dynes in the line of dip. To represent the field graphically, recourse is had to Maxwell's demonstration of the fact that the number of lines cutting unit area in different points of the field is proportional to the intensity at those points, and the numerical value of H is interpreted as the number of lines per square centimetre of a surface perpendicular to the direction of the lines. Thus at London the earth's horizontal field would be represented by drawing horizontal lines of force in the magnetic meridian, equidistant and so spaced that they cut a vertical east and west plane at the rate of .18 per square centimetre or of one line to every 5.56 + square centimetres. The positive direction is toward the north. The total field at London would be represented by lines of force in the direction of the dipping needle, equidistant and spaced so as to cut a perpendicular plane at the rate of .47 per square centimetre, or one to every 2.13 square centimeters. These lines projected intersect at the magnetic pole, and are, therefore, sensibly parallel within ordinary limits, in which case the field is said to be *uniform*.

24. Equipotential Surfaces,

Being surfaces in which no work is done in moving a unit pole, are necessarily perpendicular to the lines of force. If not, some component of the force would act, and work would be done in moving against it. Knowing the direction of the lines of force, the equipotential surfaces can be readily drawn by cutting all the lines at right angles. Like lines of force they may be drawn in any number required, but it is customary to have them represent unit difference of potential, and this requires that they should be so far apart that an erg of work is done in moving a unit pole from one to the other. The distance may be readily calculated.

$$\text{Work} = Hx,$$

but by definition work is unity

$$\therefore \ x = \frac{1}{H} \ ;$$

or the distance of the equipotential surfaces is inversely as the intensity of the field. The field may, therefore, be represented in this way as accurately as by the lines of force. Taking the case already considered, the earth's horizontal field at London would be represented by vertical surfaces, sensibly planes, extending east and west and 5.56 + centimetres apart.

25. Lines of Force due to a Single Pole.

In the case of a free pole of strength m, the number of lines of force is determined by the fact that there are m lines intersecting every square centimetre at unit distance, or that there are m lines cutting every square centimetre of a sphere of unit radius. The surface of this sphere being 4π there are in all $4\pi m$ lines of force radiating equally in all directions. The equipotential surfaces are spheres, and the radii may be found from the formula for magnetic potential, $V = \dfrac{m}{r}$, which may be here assumed to be correct. By substituting values for V differing by unity, r is found to be successively $m, \dfrac{m}{2}, \dfrac{m}{3}, \dfrac{m}{4}$, etc. As a free pole can never exist, but is always associated with another of equal but opposite polarity, any actual field is much more complex, but the cases given will illustrate the application of the principles traced, and will give clear ideas of the conventions made which underly further investigation.   ·

It is possible without changing the number of lines of force to change their distribution in the field, and as it is frequently desirable to intensify a certain part of the field, the method of doing so becomes of importance. A comparison of Figures 52 and 53 " Thompson's Electricity," shows that the distribution may be greatly changed by an arrangement of magnet poles in the field, and as iron near a magnet becomes magnetized by induction, the field is similarly affected by the introduction of iron. In this case the iron seems to gather the lines of force in the vicinity, causing a great number to pass through its substance. Iron placed near a magnet pole becomes magnetized, so that dissimilar poles are adjacent, producing · the state of affairs shown in Fig. 52. Jenkin compares this peculiarity of iron in concentrating the lines of force to that of a lens in converging rays of light. It is likewise possible to screen any part of a magnetic field from induction, by inclosing it in an iron shell. It may be easily demonstrated by experiment that if an iron ring be placed between the poles of a horse-shoe magnet, no lines of force pass through the interior of the ring, but entering at one side pass through the metal of the ring issuing on the opposite side. A magnet inside an iron sphere is independent of all outside influences. By the use of iron it is, therefore, possible either to concentrate the lines of force, or to divert them entirely from any desired part of the field.

26. Lines of Force due to a Current.

As shown in § 191, the lines of force due to a current are circles perpendicular to the current and having it for a centre. If the conductor is straight, the circles are all in parallel planes, and the equipotential surfaces are planes radiating from the conductor and each containing it. The number of these planes is such that an erg is required to move unit pole from one to the other. As shown below, the intensity of the field at unit distance is $2C$. A pole moving in a circle of unit radius having the conductor for a centre passes over a distance of 2π against a force of $2C$, doing $4\pi C$ ergs of work. The number of equipotential surfaces is, therefore, $4\pi C$. After the pole had made one revolution it would reach the equipotential surface from which it started, but having done $4\pi C$ ergs in its revolution the numerical value of the surface would now be $4\pi C$ more than before. It is impossible, therefore, to give an absolute value to an equipotential surface due to a current.

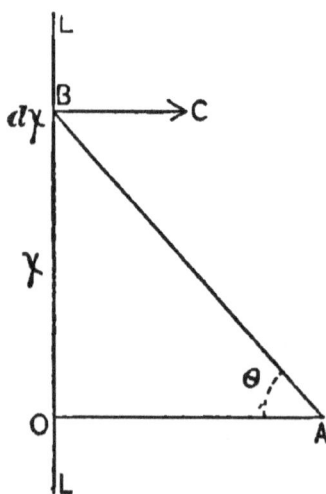

Fig. 6.

Let LL be a portion of an infinite rectilinear current, and let

BC be the force exerted by an element of length, dx, of this current $= Cdx$.   Place a unit pole at A.

The force exerted by the element dx at A is

$$dF = \frac{Cdx \cos \theta}{\overline{AB}^2}.$$

If $OA = 1$, $\overline{AB}^2 = 1 + x^2$

$$dF = \frac{Cdx . \sqrt{\dfrac{1}{1 + x^2}}}{1 + x^2} = \frac{Cdx}{(1 + x^2)^{\frac{3}{2}}}.$$

The total force $= F = C \displaystyle\int_{-\infty}^{\infty} \frac{dx}{(1 + x^2)^{\frac{3}{2}}} = C \int_{-\frac{\pi}{2}}^{\frac{\pi}{2}} \cos \theta d\theta.$

Integrating, $F = C \sin \theta \ \Big]_{-\frac{\pi}{2}}^{\frac{\pi}{2}} = 2C = H.$

To find the intensity at a distance r from the conductor we have from Note 24 that the intensity is the reciprocal of the distance between two equipotential surfaces.   There being $4\pi C$ surfaces cutting a circumference of $2\pi r$, the distance between them is $\dfrac{r}{2C}$ and hence

$$H = \frac{2C}{r}.$$

27. Magnetic Potential.

Magnetic potential has already been alluded to in Note 6.   The conception is strictly analogous to that of electrostatic potential, and the demonstration given for the formula for electrostatic potential in Note 11, is applicable to magnetic if for q a quantity of electricity, is substituted m a strength of magnet pole.   The same reasoning leads to the formula of $V = \Sigma \dfrac{m}{r}.$

Magnetic potential at a point equals the potential energy possessed by a unit north pole at that point, and is measured by the work in ergs done in bringing a unit north pole from infinity to that point.

Zero of magnetic potential exists at an infinite distance from all magnets.

Magnetic force is $\dfrac{dV}{dr}$, or the rate of change of potential per unit of distance as in Note 14.

The *difference of magnetic potential* between two points is measured by the work done in moving a unit north pole from one to the other.   Wherever work has to be done in moving a *north* pole, it would be done in resisting the motion of a *south* pole.   In all investigations of magnetic potential, force or work, a *unit north pole* must always be considered.

28. Tubes of Force.

The conception of *tubes of force* is frequently of utility. The lines of force radiating from a pole may be regarded as forming cones, and any section through a cone would cut all the lines of force.   But as the number of lines of force is proportional to the intensity, the force on all cross sections of the cone is therefore the same.   By conceiving the magnetic force to be equal throughout the cone, existing between, as well as along the lines of force, a more accurate idea of the field is attained.   It is easy to imagine a field of so slight intensity that a square centimetre would not have any lines of force passing through it.   The example of the earth's horizontal field at London, already referred to, is a case in point.   One line of force passes through every five units of area, but the magnetic forces are felt just as strongly on the four units through which the line does not pass as on that which it cuts.   By thinking, therefore, of the lines of force as *indicating* only the direction and strength of the forces which act between them as well as in them, this difficulty is overcome.

29. Intensity of Magnetization.

If a bar magnet be broken in half, instead of obtaining one piece of north and the other of south polarity, each is found to possess both and to be a perfect magnet. However far the subdivision be carried, the result is the same, and the ordinary explanation is that the magnet is an aggregation of magnetized molecules, the magnetic axes of the molecules being to a greater or less extent parallel. If this were so the north pole of one molecule would be counteracted in its magnetic effects by the south pole of the next, and the only molecules capable of exerting external magnetic effects would be those on the surface, and the effect is exactly the same as would be produced by a distribution of a magnetic matter or fluid, or avoiding the idea of a fluid, a distribution of magnetism over the surface of the magnet. The amount of magnetism per unit area is called the *magnetic density*. If the magnetism is regarded as being uniformly distributed throughout the mass of the magnet, the quotient of the magnetic current by the volume is called the *intensity of magnetization*.

Let ρ be the magnetic density, a the cross section and l the length of the magnet. Then the strength of pole is

$$m = \rho\, a$$

$$m\, l = \rho\, a\, l$$

$$\therefore \rho = \frac{ml}{al}$$

$$= \frac{\text{Magnetic moment}}{\text{Volume}}.$$

The intensity of magnetization and magnetic density are therefore practically the same, the one presupposing a

uniform distribution of magnetism throughout the mass,
the other a surface distribution.

If the magnetism is due to the bar being situated in a
magnetic field, the intensity of magnetization is equal to
$k\,H$, k being what is called a "coefficient of magnetization." A few values of k are given in ¿ 340 (" Thompson "). Assuming Barlow's value for iron 32.8, the formula,
intensity of magnetization $= k\,H$ indicates that the intensity of magnetization is dependent only on the intensity
of the field ; but there is found to be a limiting value of
magnetization which cannot be exceeded, however powerful the field is. This is stated to be for iron 1390 (p. 269,
" Thompson "), and the strongest field that could be utilized in magnetizing iron is therefore $\dfrac{1390}{32.8} = 42.4.$ The
value of this coefficient is, however, uncertain, and appears
to be much less at a high intensity of magnetization than
at a low

30. Solenoidal Magnets.

A filament of magnetic matter so magnetized that its
strength is the same at every cross section is called a
magnetic solenoid. A long thin bar magnet uniformly
magnetized is called a *solenoidal magnet*, or simply a
solenoid, in distinction to a magnetic shell. The name
solenoid is also applied to a helix through which a current
passes. (See Note 42.) As the magnet poles are points at
which the magnetism of the magnet may be supposed to
be concentrated, and from which magnetic forces act, the
potential of any point near the magnet is determined by its
distance from the two poles, or $V = \Sigma \dfrac{m}{r} = m\left(\dfrac{1}{r} - \dfrac{1}{r'}\right).$
The exact position of the poles is difficult to determine,
but is stated in ¿ 122 to be in long thin steel magnets

about $\frac{1}{10}$ of the distance from the end.   If the poles are bent to meet, forming a ring, $r = r'$ for all external points, and there is therefore no potential due to a magnetized ring.

31. Potential due to a Magnetic Shell.

As defined in ¿ 107 (" Thompson ") a magnetic shell is a thin sheet so magnetized that the two sides of the sheet have opposite kinds of magnetism.   The demonstration and use of the expression for the potential due to a magnetic shell requires a preliminary definition of a solid angle, and of the method of measuring it.   The solid angle subtended at any point by a closed curve or surface *is measured by the area of a sphere of unit radius described from the point as a centre, intercepted by lines drawn from all parts of the curve to the point.*   (See Fig. 64, " Thompson.")   As the areas of similar surfaces on spheres are as the square of the radii the solid angle,

ω = area on unit sphere,

$$= \frac{\text{area on sphere of radius } r}{r^2},$$

$$= \frac{\text{area on sphere of radius } r_1}{r_1^2}.$$

To compute the solid angle.

When the closed curve is circular and the point is in its axis it is necessary only to compute the area of a zone of one base on a sphere of unit radius.   The area of the zone formed by the revolution of AD around AO as an axis is (see Chauvenet's Geometry, Book IX., Prop. X. Cor. III)

$$Ad \times 2 \pi OA.$$

But $Ad = AO - dO = r - r \cos \theta$

$$\therefore \text{ Area} = 2\pi r (r - r \cos \theta);$$

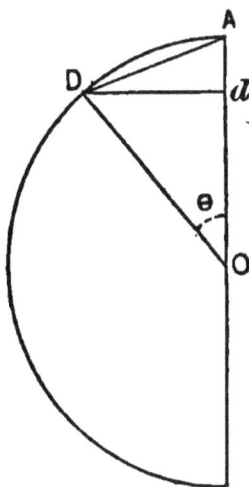

Fig. 7.

3

but if $r = 1$, area is solid angle,

$\therefore \; \omega = 2\pi \, (1 - \cos \theta)$.

To calculate the potential.

Let r_1 and r_2 be the distances from the point D to the faces of the small element ds, β be the angle between ds and its projection

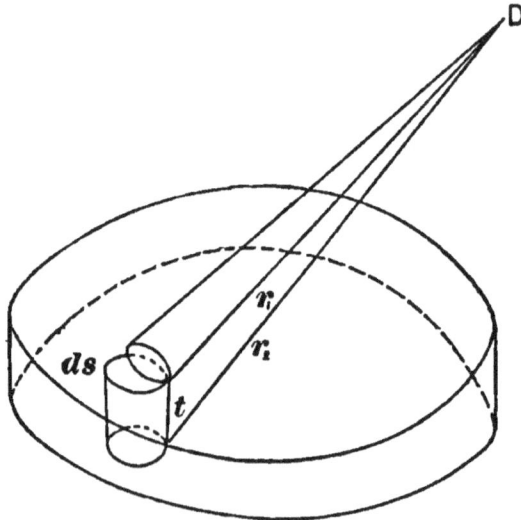

Fig. 8.

perpendicular to r_1 and ρ be the magnetic density. The strength of the shell i being the product of the density and thickness,

$$\cdot \quad i = \rho t \; \therefore \; \rho = \frac{i}{t}.$$

The quantity of magnetism on the small element ds is

$$dm = ds.\rho = ds.\frac{i}{t} \; . \quad . \quad . \quad . \quad . \quad . \quad (1)$$

The potential at D due to ds is

$$dV = dm \left(\frac{1}{r_1} - \frac{1}{r_2} \right) = \frac{dm}{r^2} \, (r_2 - r_1) \; . \quad . \quad . \quad (2)$$

But $\cos \beta = \frac{r_2 - r_1}{l}$. Substituting this value and that of dm in (2)

$$dV = \frac{ids}{r^2} \cos \beta \quad \cdots \cdots \quad (3)$$

But as ds is an infinitely small element, its plane projection $ds \cos \beta$ perpendicular to r_1 is sensibly equal to the area on a sphere of radius r. The solid angle, therefore, subtended at D by the element ds is

$$d\omega = \frac{ds. \cos \beta}{r^2}$$

Substituting in (3)
$$dV = d\omega i$$
$$V = \omega i.$$

32. Equipotential Surfaces and Lines of Force of a Magnetic Shell.

The computation of the solid angle is, as shown, simple when the point D is on the axis normal to the shell at its centre, but when D is oblique the area becomes an ellipse on a spherical surface of unit radius, and the calculation is extremely difficult. From the formula, however, for magnetic potential a few relations are readily deduced. As at all points where the shell subtends the same solid angle the potential is the same, any equipotential surface is evidently most remote from the shell on the axis normal to its centre. As the point of view becomes oblique it must approach the shell that the solid angle may be the same, and at all points in the plane of the shell the solid angle, and consequently the potential, are zero. The general form of the equipotential surfaces is, therefore, that of deep bowls concave to the shell, and most remote from it on its perpendicular axis. At a point close to the shell the solid angle is a hemisphere or 2π and the potential $2\pi i$. On the opposite side the potential is $-2\pi i$, or $4\pi i$ ergs must be expended in moving unit pole from one side of the shell to the other. If the equipotential surfaces indicate unit difference of potential there are, therefore, $4\pi i$ surfaces. From the equipotential surfaces the direction of the lines of force may be traced, as they start from the side of the shell having north polarity and curve so as to cut each surface at right angles, finally entering the south pole of the shell at right angles.

33. Work Done in Moving Pole near Shell.

Potential being measured by the work done on unit pole in bringing it up to a point from an infinite distance, the work done on a pole of strength m is $m\omega i$. It is possible under the conventions made as to the number of lines of force to express this in another way. As already shown, the number of lines of force given off by a pole of strength m is $4\pi m$, but as these radiate in all directions, they are given off throughout a solid angle 4π subtended at the centre of a sphere. Through any solid angle ω the number of lines is, therefore, $m\omega$. Calling this number N, the above expression becomes Ni, or the work done in bringing a pole up to a position near a magnet shell is measured by the product of the strength of the shell and the number of lines of force of the pole cut by the shell. This is evidently a measure of the work done either in bringing the pole up to the shell or the shell to the pole, and is, therefore, sometimes called the *mutual potential* of the pole and shell. The work done in bringing the pole from infinity to a point where it intercepts N_1 lines is $N_1 i$. If now it be moved to another in which it intercepts N_2 lines, the work done between the points is

$$\text{Work done} = i\,(N_2 - N_1).$$

The difference of potential between the points is

$$V_2 - V_1 = i\,(\omega_2 - \omega_1)\,; \quad \text{or } \frac{i\,(N_2 - N_1)}{m}\,;$$

magnetic potential being the quotient of work done in moving pole, by the strength of pole.

The work done may be either positive or negative and the above expression may, therefore, have either sign. If N_1 > N_2 the work done in passing from N_1 to N_2 is negative, or the shell tends to move in such a direction as to include a minimum number of lines of force. As these pass in the positive direction, exactly the same relation is expressed by saying that a magnetic shell in a field tends to place itself so as to enclose the maximum number of negative lines of force. If the north pole of a magnet shell is brought up to the north pole of a magnet, this relation is readily seen, as the shell will be repelled into a position in which it will enclose as few lines of force taken in the positive direction as is possible. If the same face be approached to a south pole, it is attracted and moves into a position in which the maximum number of lines cut the shell in the negative direction.

34. Equivalent Magnetic Shells.

The relations deduced for magnetic shells are of great service, as they are applicable to the case of a voltaic circuit in a magnetic field. If a wire carrying a current be looped into a circle, the lines of force which ordinarily encircle the conductor combine to act in the same direction on a pole at a distance from the circuit. Thus in Fig. 86 (" Thompson "), it is seen that all the lines of force due to the current pass in the same direction through the plane of the circuit as do those of a magnetic shell. A closed voltaic circuit in a magnetic field, as may be readily shown by experiment, is acted upon as a magnet would be. It is found that the magnetic effects of the north pole of a magnet are identical in nature with those of a circuit, in which the current flows in a direction opposite to that in which the hands of a watch move. This direction is known as the negative direction of the current, and the magnetic effects of a positive pole and of a negative current are, therefore, similar. Looking at the other side of the loop, the current would appear to pass in the direction in which the hands of a watch move, or in the positive direction ; but if a north pole be approached to the circuit from the side on which the current appears to have this direction it is attracted, showing that a positive current produces magnetic effects similar to those of a negative pole. As the direction in which a north pole moves shows the direction of the lines of force, it is seen from the above that the lines of force enter that face of the plane of the circuit in which the current appears to move " with the sun," or in the positive direction, and emerge from the other face. As it is a matter of great importance to be able to connect the direction of the lines of force with that of the current to which they are due, several rules have been given, one of the best of which is the comparison between the direction of rotation of a corkscrew and that of the motion of its point. If the wrist be rotated in the right-handed direction, the point advances ; and considering the motion of the wrist to be that of the current, the movement of the point corresponds to that of a north pole, and indicates the direction of the lines of force. This relation is said to be that of " right-handed cyclical order," and the direction of the current and of the lines of force are spoken of as being thus related.

The magnetic action of a voltaic circuit is found to depend upon the strength of the current, and on the area of the enclosed surface. It is, therefore, evident that for every closed circuit, a magnetic shell whose edges coincided in position with the circuit could be substituted, if a certain relation were established between the units measuring the strength of the current and the strength of the shell. This relation is that expressed in the definition in ₰ 195. The absolute electromagnetic unit of current is that current which in passing through a conductor one centimetre long, bent so as to be in all parts distant one centimetre from a unit pole, acts on the pole with a force of one dyne. By this definition unit current produces unit field at unit distance. But so does a shell or pole of unit strength. By expressing, therefore, the current in these units, the magnetic effect of the current is the same as that due to a magnetic shell whose edges coincide with the circuit, and whose strength is equal to that of the current. This is called an equivalent magnetic shell, and all relations hitherto traced for the shell are now applicable to the closed circuit.

35. Potential due to a Closed Voltaic Circuit.

The potential due to a current at a point is therefore $C\omega$, where C is the current measured in absolute units, and ω is the solid angle subtended by the circuit at the point. As, however, a positive current produces the same magnetic effects as a negative pole, the sign of the potential is always the opposite of that of the current, or

$$V = - C\omega.$$

The difference of potential between two points is

$$V_2 - V_1 = - C(\omega_2 - \omega_1).$$

The magnetic force due to a current at a distance x is (Note 14)

$$\frac{dV}{dx} = - C. \, d\omega.$$

The potential is $2\pi C$ on one side of the circuit and $- 2\pi C$ on the other, changing by $4\pi C$. Hence there are $4\pi C$ equipotential surfaces.

36. Work Done in Moving a Circuit Near a Pole.

This is a problem of the greatest importance, as it underlies the action of the dynamo machine. As already traced (Note 33), the work done in moving a magnetic shell near a pole, or conversely the pole near the shell, is

$$\text{Work done} = m\omega i = Ni.$$

Similarly the work done in moving a closed circuit near a pole is

$$\text{Work done} = - m\omega C = - NC,$$

N being the number of lines of force of the pole passing through the circuit. If the circuit be brought up from an infinite distance to a point where it intersects N_1 lines due to the pole the work is $- N_1 C$. If now moved still farther so that it intersects a greater number, N_2, the work done between the points is

$$\text{Work} = - C(N_2 - N_1),$$

and

$$\text{Difference of potential} = - \frac{C(N_2 - N_1)}{m} = - C(\omega_2 - \omega_1).$$

If $N_2 > N_1$ the work is negative, and the circuit tends to move therefore in such a manner as to make the number of lines enclosed a maximum. If a circuit be placed in a magnetic field so that the lines of the field while parallel to those of the current pass in the opposite direction, the circuit will, if free, first turn to bring the lines in the same direction, and will then move to make the number enclosed a maximum (Fig. 87, " Thompson ").

It may be useful to obtain an expression for the work done in this last case, as an understanding of the theory will assist in the comprehension of the working of electric-motors. Imagine a closed circuit placed in a uniform field. If the circuit be moved parallel to itself, the number of lines enclosed is constant, and consequently no work is done in whatever direction the circuit be moved. If, however, the coil be rotated on an axis in its plane, it will enclose a varying number. If θ be the angle between the normal to the plane of the coil and the direction of the lines of force of the field, and the number of lines passing through the coil when its plane is perpendicular to them be N, the number enclosed when at an angle θ is $N \cos \theta$. If the angle be now changed to θ' the number enclosed is $N \cos \theta'$, and the work done in passing from one position to the other is

$$\text{Work} = - C (N \cos \theta' - N \cos \theta).$$

Suppose that the coil be rotated. The work done is easily calculated :

In the first quadrant,
$$\text{work} = - C (N \cos 90° - N \cos 0°) = CN;$$
in the second quadrant,
$$\text{work} = - C (N \cos 180° - N \cos 90°) = CN;$$
in the third quadrant,
$$\text{work} = - C (N \cos 270° - N \cos 180°) = - CN;$$
and in the fourth quadrant,
$$\text{work} = - C (N \cos 0° - N \cos 270°) = - CN.$$

In the first half of the revolution, therefore, work equal to $2CN$ has to be done in order to move the coil, but in the latter half the coil will do the same amount of work. The potential energy of the coil is, therefore, greatest when $0 = 180°$, or when the lines of force of the field are parallel to those of the coil, but in the opposite direction, and if the coil be then left free to move, it will rotate to make $6 = 0°$, doing work equal to $2CN$, and then requiring work to be done on it to cause further movement. On the supposition already stated that the field is uniform, $N = HA$, H being the intensity of the field and A the area of the coil. The work done by the coil in rotating through two quadrants may then be expressed as $2CHA$, this also measuring the work done on the coil in the other two quadrants.

As a résumé of the above, we have the rule, *that a movable circuit in a magnetic field tends to place itself so as to enclose the maximum number of lines of force in right-handed cyclical order.*

37. To Calculate the Intensity of the Field due to a Voltaic Circuit.

The force acting on unit pole, or the intensity of the field, is by Note 35 the rate of change of potential per unit of length. The intensity of field at a distance x is

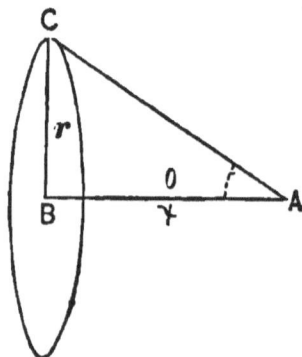
Fig. 9.

$$H = \frac{dV}{dx} = - C d\omega.$$

The difficulty of calculating the value of ω makes the general solution extremely complicated. It is, however, easy to calculate the intensity of the field at any point on the axis of the circuit, as in that case $\omega = 2\pi (1 - \cos \theta)$.

Let $x =$ the distance of the point A from the circuit.
 $r =$ radius of the coil.

Then $dV = -2\pi C.d\,(1 - \cos\theta)$ $\quad \cos\theta = \dfrac{x}{(x^2 + r^2)^{\frac{1}{2}}}$.

$$\therefore \frac{dV}{dx} = -2\pi C\left(\frac{1}{(x^2 + r^2)^{\frac{1}{2}}} - \frac{x^2}{(x^2 + r^2)^{\frac{3}{2}}}\right)$$

$$= -\frac{2\pi C r^2}{(x^2 + r^2)^{\frac{3}{2}}}.$$

At the centre of the circle the force is a maximum, and is $\dfrac{2\pi C r^2}{(r^2)^{\frac{3}{2}}} = \dfrac{2\pi C}{r}$ as in § 195. The $-$ sign shows that with a positive current the force is one of attraction.

38. Position of Equilibrium of a Circuit and Magnet.

Consider the magnet as composed of two poles of strength m and $-m$ connected rigidly. The formula for the force in the field may be written

$$\frac{2\pi C r^2}{(x^2 + r^2)^{\frac{3}{2}}} = \frac{2\pi C}{r}\sin^3\theta.$$

The forces acting on the two poles are

$$\frac{2\pi C}{r}\sin^3\theta \quad \text{and} \quad -\frac{2\pi C}{r}\sin^3\theta'.$$

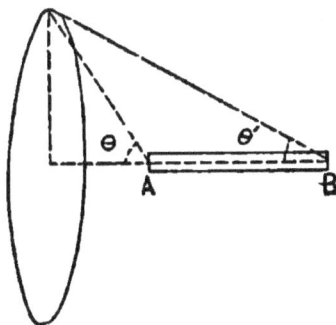

Fig. 10.

The resultant force is

$$\frac{2\pi C}{r}(\sin^3\theta - \sin^3\theta').$$

This is zero when $\theta = \theta'$, or when the centre of the magnet is at the centre of the coil. In any other position there will be a force acting and the equilibrium will be unstable. If, therefore, either the coil or magnet is free to move, the coil will place itself so that the middle of the magnet is at its centre. (See Fig. 87, " Thompson.")

39. Mutual Potential of two Circuits (§ 320).

The work done in bringing one circuit up to another, or the " mutual potential " of the two circuits is, as given in § 320, $- cc'. \dfrac{\cos \varepsilon}{r} ss'$. This expression is one of great theoretical importance, but its derivation is difficult and out of place in an elementary treatise.

The work done in moving a circuit near a pole or in a field has already been shown to be $- NC$, and it is obviously immaterial whether the lines of force N are due to a pole, a magnetic shell or another circuit. Suppose two circuits A and B, carrying currents of strength C and C' and let N_1 be the number of lines of force due to A enclosed by B, and N_2 the number due to B enclosed by A. If B is moved out of the field caused by A the work done is $N_1 C$. If A is now moved so as to resume its former relative position to B the work done is $- N_2 C'$. The coils are now in the same relative position as at first and if there are no external magnetic forces, no work can have been done in moving the system. Hence

$$N_1 C - N_2 C = 0.$$

If the current in each is of unit strength

$$N_1 = N_2,$$

or each encloses the same number of the other's lines of force. Returning to the expression for the work done. $CC' \dfrac{\cos \varepsilon}{r} ss'$, and making C and C' each unity, the number of lines enclosed by each is $\dfrac{\cos \varepsilon}{r} ss'$, and this number may be represented by the single symbol M and is dependent only on the position and areas of the two coils.

Let the planes of the two circuits be parallel, and the current flowing in the same direction in each, that in which the hands of a watch move. The negative sign of the formula shows that the circuits attract each other, and this is also evident from Maxwell's rule that a voltaic circuit free to move always places itself so as to enclose the greatest possible number of lines of force. The nearer the circuits, the greater the value of M, and if they become coincident M would be $\frac{ss'}{0}$, or infinite. As, however, r will always have a finite value, the maximum value of M exists when the coils are touching, or r is a minimum. As the coils tend to approach or to diminish, r, the "coefficient of mutual potential" M, always tends to a maximum. This quantity is hereafter referred to (Note 63) as the "coefficient of mutual induction."

40. Conversion of Units (§ 324).

The use of the dimensions of units in passing from one system to another has been illustrated in Note 19. In electrical calculations, the most frequent change to be made is that from the C.G.S. system to the British units based on the foot, grain, and second. The ratios between these units are shown in the following table from Jenkin's "Electricity."

		Symbol.	Number of C.G.S. Units in one British Unit (A).	Logarithm of A.	Logarithm of B.	Number of British Units in one C.G.S. Unit (B).
Fundamental.	Mass	M	0.0647989	$\bar{2}$.8115678	$\bar{1}$.1884321	15.43235
	Length	L	30.47955	1.4840071	2.5159929	0.03280899
	Time...............	T	1.			1.
Derived.	Force..............	F	1.97504	0.2955749	$\bar{1}$.7044250	0.50632
	Work	W	60.198	1.7795820	2.2204179	0.01661185
Electro-Static.	Quantity	q	42.8346	1.6317949	$\bar{2}$.3682051	0.0233456
	Current............	i or C	42.8346	1.6317949	$\bar{2}$.3682051	0.0233456
	Potential..........	V	1.40536	0.1477874	$\bar{1}$.8522125	0.711561
	Resistance.........	R	0.03280899	$\bar{2}$.5159929	1.4840071	30.47945
	Capacity...........	C	30.47945	1.4840071	$\bar{2}$.5159929	0.03280899
Magnetic.	Strength of Pole....	m	42.8346	1.6317949	$\bar{2}$.3682051	0.0233456
	Magnetic Potential.	V	1.40536	0.1477874	$\bar{1}$.8522125	0.711561
	Intensity of Field...	H	0.0461085	$\bar{2}$.6637804	1.3362196	21.6880
Electro-Magnetic.	Current.	i or C	1.40536	0.1477874	$\bar{1}$.8522125	0.711561
	Quantity	Q	1.40536	0.1477874	$\bar{1}$.8522125	0.711561
	Potential... ...... } Electromotive F'ce {	V E	42.8346	1.6317949	$\bar{2}$.3682051	0.0233456
	Resistance.........	R	30.47945	1.4840071	$\bar{2}$.5159929	0.03280899
	Capacity...........	C	0.03280899	2.5159929	1.4840071	30.47945

The following table, showing the relations between the practical units in common use, may be convenient for reference :

1 metre = 39.37043 inches = 3.28087 feet.

1 kilogramme = 2.20462 avoirdupois pounds.

1 kilogrammetre, or 1 kilogramme raised one metre per second = 7.23307 foot-pounds per second.

1 Force de cheval or French H. P. = 75 kilogrammetres per second.

1 English H. P. = 33000 foot-pounds per minute.

$$= 550 \quad `` \quad `` \quad `` \text{ second.}$$
$$= 76.04 \text{ kilogrammetres per second.}$$
$$= 1.014 \text{ force de cheval.}$$

1 gramme = 981 dynes (980.868 at Paris).

1 gramme centimetre, or 1 gramme raised one centimetre in a second = 981 ergs.

1 pound avoirdupois = 4.45 × 10⁵ dynes nearly.

1 foot-pound per second = 1.356 × 10⁷ ergs nearly.

1 Volt = 10⁸ absolute electromagnetic units of potential.

1 Ohm = 10⁹ `` electromagnetic units of resistance.

Practical Ohm = .9895 × 10⁹ absolute units (Lord Rayleigh).

1 Ampère = $\frac{1}{10}$ of an absolute electromagnetic unit of current.

1 Coulomb = $\frac{1}{10}$ of an absolute electromagnetic unit of quantity.

$$= 1 \text{ ampère per second.}$$

1 Farad = 10⁻⁹ absolute electromagnetic units of capacity.

1 Watt (see Note 55).

$$= 10^7 \text{ ergs} = .7373 \text{ foot-pounds.}$$
$$= \tfrac{1}{746} \text{ English H. P.}$$

1 thermal unit = 1 gramme of water raised 1° C.

1 Joule = 10⁷ ergs = .24 thermal unit (See Note 55).

Mechanical equivalent of heat = 772 foot-pounds 1° F.

$$= 1390 `` \quad `` \quad 1° \text{ C.}$$

Same in metric system = 424 kilogrammetres 1° C.

$$= 42400 \text{ gramme-centimetres } 1° \text{ C.}$$
$$= 4.16 \times 10^7 \text{ ergs.}$$

1 Siemens unit = .9536 practical ohm.

1 Jacobi = current evolving 1 $\overline{Cm}^3$. of mixed gas per minute at 0° and 760 mm.

= .095 ampères.

At any place the weight of the gramme is equal to g dynes. The value of g for any latitude may be found approximately from the formula

$$g = 980.6056 - 2.5028 \cos 2\lambda - .0000003h,$$

λ being the latitude and h the height above sea level. The limiting values of g are 978.1 at the equator and 983.1 at the poles.

41. Determination of the Horizontal Component of the Earth's Magnetism (§ 325 a)

The time of vibration of a particle acted on by a constant force is $t = \pi \dfrac{1}{\sqrt{\mu}}$, t being the time of a half or simple vibration, and μ the acceleration. The latter is in any case equal to the moment of the impressed forces divided by the moment of inertia. When the arcs of vibration are small this may be applied to a magnet oscillating in a uniform field and the time of a complete or double oscillation of a magnet is therefore

$$t = 2\pi \sqrt{\dfrac{1}{\dfrac{mlH}{K}}} = 2\pi \sqrt{\dfrac{K}{MH}} \quad \cdot \quad \cdot \quad \cdot \quad (1)$$

K being the moment of inertia. M the magnetic moment and H the horizontal intensity.

$A.$—To make the observation, a magnet is allowed to oscillate and the time of vibration is determined as accurately as possible. This is best done by determining the approximate time of one oscillation, and allowing the magnet to oscillate a known time. Dividing this time by the approximate time of one oscillation gives the approximate number of oscillations. Taking the nearest whole number to this, and dividing the whole time by it gives the exact time of one vibration. If the approximate number of vibrations fell midway between two whole numbers, the observation would have to be repeated until it was known with certainty how many oscillations had been made in the observed time. The oscillations must be of small amplitude, and in very-exact observations must be reduced to an infinitely small arc. If possible, the magnet should be supported by a single fibre to avoid torsion, but if a wire has to be used, the torsion must be allowed for. If the magnet is of simple form, either bar or cylindrical, the value of K may be determined from the formulas in § 325a, but if these do not apply, K may be determined by observation of the time of vibration of the magnet, and of the time, t_1, when its moment of inertia is increased by the addition of a weight of known moment K^1.

From (1) the value of MH is $\dfrac{4\pi^2 K}{t_1^2}$ . . . (2)

When the weight is added, $MH = \dfrac{4\pi^2 (K + K^1)}{t_1^2}$

∴ $K : K + K^1 :: t^2 : t_1^2$ or $K = \dfrac{K^1 t^2}{t_1^2 - t^2}.$

Knowing K it is possible to compute MH from (1).

In the case of the oscillating magnet, the magnet and earth acted mutually on each other. In order to obtain the ratio $\dfrac{M}{H}$ the force of the magnet must act against that

4

of the earth, and this is done by measuring the deflection from the magnetic meridian that the magnet will cause in a small magnetic needle near it. NS is the magnet the time of whose oscillation has been determined, and it is placed at right angles to the magnetic meridian, so that its centre is due north or south of the small needle at O. Let r be the distance of either N or S from O, and m be the strength of N and S. Then the force exerted by S on a south pole m' at O is

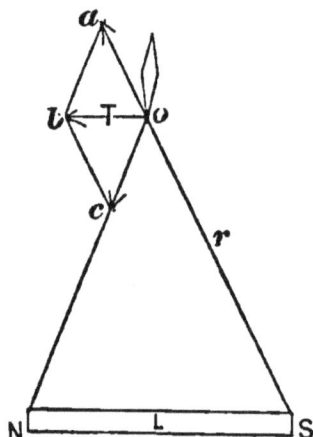

Fig. 11.

$\dfrac{mm'}{r^2}$ in the direction Oa, which may be taken to represent it in magnitude and direction. Similarly Oc would represent the force of attraction of the pole N. The resultant force is Ob.

From similar triangles

$$Oa : Ob : : So : NS ;$$

or calling Ob, T, and NS, L,

$$\frac{mm'}{r^2} : T : : r : L \quad \therefore T = \frac{mm'L}{r^3}.$$

Let $M' = m'l'$, the moment of the small magnet, and $M = mL$ the moment of the large one. Then the couples acting on the small needle when it has a permanent deflection θ are ;

to deflect, $Tl' \cos \theta = \dfrac{mm'Ll'}{r^3} \cos \theta = \dfrac{MM'}{r^3} \cos \theta ;$

to retain in the meridian, $m'l'H \sin \theta = M'H \sin \theta.$

Equating these moments and reducing

$$\frac{M}{H} = r^3 \tan \theta. \quad \ldots \ldots \quad (3)$$

By combining (2) and (3)

$$H = \frac{2 \pi}{t} \sqrt{\frac{K}{r^3 \tan \theta}}. \quad \ldots \ldots \quad (4)$$

B.—In the Kew Magnetometer the deflecting magnet is placed east or west of the small magnet instead of north or south, and the formula is different.

Let *r* be the distance between the pole *s* and the centre

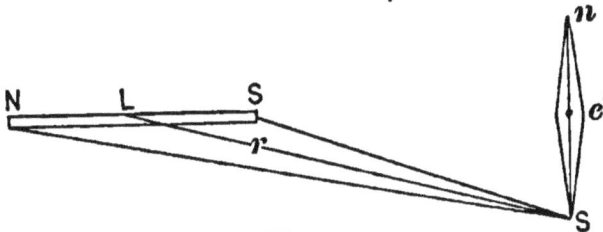

Fig. 12.

of the deflecting magnet, *L* be the length of the deflecting magnet, *m* the strength of pole of the deflecting and *m'* that of the deflected magnet, and *l'* the length of the latter. The force of repulsion exerted by *S* on *s* is $\dfrac{mm'}{\left(r - \dfrac{L}{2}\right)^2}$, *r* being great in comparison with *L*. The force of attraction of *N* on *s* is $\dfrac{m\,m'}{\left(r + \dfrac{L}{2}\right)^2}$. These forces may be considered as acting in the same line but in opposite directions, and the resultant force acting on *s* is

$$F = mm' \left(\frac{1}{\left(r - \dfrac{L}{2}\right)^2} - \frac{1}{\left(r + \dfrac{L}{2}\right)^2} \right)$$

$$= mm' \left(\frac{2Lr}{\left(r^2 - \frac{L^2}{4}\right)^2} \right) = 2Mm' \left(\frac{r}{\left(r^2 - \frac{L^2}{4}\right)^2} \right). \quad (5)$$

The moment of the deflecting couple on the small magnet is

$$Fl' \cos \theta = 2MM' \cdot \frac{r}{\left(r^2 - \frac{L^2}{4}\right)^2} \cos \theta ,$$

or since as above, L is small in comparison with r, its square and higher powers may be neglected, and the moment of the couple is

$$2MM' \cdot \frac{r}{\left(r^2 - \frac{L^2}{4}\right)^2} \cos \theta = \frac{2MM'}{r^3} \cos \theta .$$

The moment of the couple tending to retain the small magnet in the meridian is

$$M'H \sin \theta.$$

Equating and reducing

$$\frac{M}{H} = \tfrac{1}{2} r^3 \tan \theta . \quad . \quad . \quad . \quad . \quad (6)$$

C.—This derivation contains many assumptions. A more rigorously correct formula is that given by Kohlrausch,

$$\frac{M}{H} = \tfrac{1}{2} \frac{r^3 \tan \theta - r_1^3 \tan \theta'}{r^2 - r_1^2} .$$

In which r and r_1 are the distances between the centres of the magnets in two successive positions, and θ and θ' the corresponding angles of deflection. The deflecting magnet is placed east or west of the deflected needle as in the last case.

The formula of (6) is the one generally used in work with the Kew Magnetometer, but is true only when r is large in comparison with L. The more accurate formula can be readily derived. From equation (5):

$$F = 2Mm' \left(\frac{r}{\left(r^2 - \frac{L^2}{4} \right)^2} \right) = \frac{2Mm'}{r^3} \left(1 - \frac{L^2}{4r^2} \right)^{-2}$$

$$= \frac{2Mm'}{r^3} \left(1 + \frac{1}{2} \frac{L^2}{r^2} \cdot \cdots \cdot \right)$$

The moment of the deflecting couple is

$$Fl' \cos \theta = \frac{2MM'}{r^3} \left(1 + \frac{1}{2} \cdot \frac{L^2}{r^2} \right) \cos \theta .$$

Equating this with $M'H \sin \theta$, the moment tending to retain the couple in the meridian, and reducing,

$$\tan \theta = \frac{2M}{r^3 H} \left(1 + \frac{1}{2} \cdot \frac{L^2}{r^2} \right) . \quad \cdots \quad (7)$$

By repeating the observation by placing the deflecting magnet so that its centre is at a distance of r_1 from the needle, a new deflection, θ', is obtained, and

$$\tan \theta' = \frac{2M}{r_1^3 H} \left(1 + \frac{1}{2} \cdot \frac{L^2}{r_1^2} \right) . \quad \cdots \quad (8)$$

Multiplying (7) by r^5 and (8) by r_1^5, and subtracting the latter from the former,

$$r^5 \tan \theta - r_1^5 \tan \theta' = 2 \cdot \frac{M}{H} (r^2 - r_1^2) ;$$

or,

$$\frac{M}{H} = \frac{1}{2} \cdot \frac{r^5 \tan \theta - r_1^5 \tan \theta'}{r^2 - r_1^2} .$$

IV. MEASUREMENTS AND FORMULÆ.

42. Solenoids (§ 327). Ampère's Theory of Magnetism (§ 338).

As stated in § 327, a spiral coil of wire through which a current passes is called a solenoid. The definition has already been given as that of a magnetic filament uniformly magnetized, and as a spiral coil carrying a current exerts the same forces, and is similarly acted upon in a magnetic field, it is called by the same name. Theoretically the turns of the coil should be exactly parallel, and at right angles to the longitudinal axis, but as this is impossible the ends of the helix are brought in through the coil from each end to the centre as in Fig. 116. The current flowing in these branches exerts an effect equal and opposite to that due to the longitudinal component of the spiral, and the resultant effect of a solenoid thus constructed is that of a number of parallel turns only, or of the theoretical solenoid.

If the helix be free to move it will, when a current is passed through it, move so as to include the maximum number of lines of force in the field. In the earth's field, it therefore assumes the same position as the dipping needle. The end pointing north is called the north pole of the solenoid, as with the ordinary bar magnet. Let a solenoid be suspended so as to move freely, and a magnet be brought near it so that the north poles are nearest. The solenoid will be repelled, but if after repulsion, the south pole of the magnet is brought up, the north pole of the solenoid is attracted. Magnetic forces are evidently acting between the coil through which a current is flowing and the piece of steel which we call a magnet. Many other similar effects have been alluded to, and they suggested to Ampère a theory of magnetism, which explains very many peculiar relations and is of great practical utility. He conceived that a magnet was composed of a great number of molecules, around each of which flowed an electric current in a constant direction. In an ordinary unmagnetized bar of steel, these currents lie in all possible planes, so that their resultant magnetic effect is zero. If, however, the bar be magnetized, the energy expended in so doing operates to turn the molecules so that the currents are now parallel. In looking at the end of a magnet, the molecular currents would, as shown in the figure, counteract each other in the substance of the magnet, but the current on the outer edge of the outer row of molecules being unbalanced would cause a resultant current on the surface in the direction opposite to the motion of the hands of a clock. In

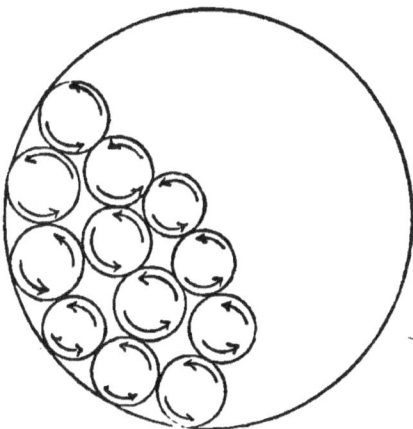

Fig. 13.

a solenoid a current flows in this direction when looking at the north pole of the solenoid, and the figure therefore shows the theoretical condition of the north pole of the magnet. Looking at the other face the currents appear to flow in the opposite direction. From these suppositions follows the rule given on p. 284. The theory explains many peculiar effects, such as those referred to in § 112 and § 113, but is no more than a theory. It seems unquestionable that the process of magnetization is attended by molecular movements, but it is not proved that magnetism is due to molecular currents. Prof. Hughes has recently experimented on tempered steel, using an ingenious modification of the induction balance, and asserts his belief that each molecule possesses magnetic polarity. In tempered steel the molecules are comparatively fixed, whereas in soft iron they possess considerable freedom of movement. He thinks the process of magnetization is merely one of molecular movement, by which the similar poles of the molecules are brought facing the same way. This final position is retained by the steel but lost by the iron.

Ampère investigated the mutual action of magnets and currents by the theory of action at a distance between currents, but if in Fig. 13 the current in each molecule is supposed to have one line of force, the aggregation of molecules would produce the large number of lines all passing in the same direction that the magnet is found to possess, and the theory of lines of force due to Faraday and Maxwell explains all magnetic phenomena as well if not better than the method of Ampère.

43. Best Arrangement of Cells (§ 351).

From general formula, letting w be the total number of cells, $w = m\,n$

$$C = \frac{mE}{\dfrac{mr}{n} + R} = \frac{E}{\dfrac{r}{n} + \dfrac{R}{m}} = \frac{E}{\dfrac{mr}{w} + \dfrac{R}{m}} \;*\;.\;\;.\;\;(1)$$

---

* Proof without calculus.

In equation (1) $\dfrac{mr}{w} + \dfrac{R}{m} = 2\sqrt{\dfrac{Rr}{w}} + \left(\sqrt{\dfrac{mr}{w}} - \sqrt{\dfrac{R}{m}}\right)^2$

This is a minimum when the square it contains is zero, or when

$$\sqrt{\frac{mr}{w}} = \sqrt{\frac{R}{m}}.$$

In that case $R = \dfrac{m^2 r}{w} = \dfrac{mr}{n}$ as before.

This is a maximum when the denominator is a minimum. Differentiating with regard to m and making first derivative equal zero.

$$\frac{du}{dm} = \frac{r}{w} - \frac{R}{m^2} = 0 \;\; \therefore \;\; \frac{r}{w} = \frac{R}{m^2} \text{ or } R = \frac{m^2 r}{w} = \frac{mr}{n}.$$

But $\dfrac{mr}{n}$ is the internal resistance of the battery. Hence the rule, that the best arrangement is secured when the internal resistance of the battery equals the external resistance in circuit.

44. Long and Short Coil Galvanometers (§ 352).

In the use of a galvanometer it is desirable that it should produce a readable deflection without greatly reducing the current. If the current is large a single turn of wire will cause a sufficiently strong field, but if the current is small, it is necessary to multiply its effect by passing it through many turns in order to obtain a good deflection.

By Note 37 the field at the centre of the coil is $\dfrac{2\pi Cn}{r}$ in a galvanometer having n, and $\dfrac{2\pi C}{r}$ in another having only one turn around its needle. The resistance of the first will be nearly n times that of the second. Let all the resistance in the circuit external to the galvanometer be r, and g be the resistance of the galvanometer, and E the E. M. F., supposed constant. Then

$$C = \frac{E}{r + g} \quad \text{and} \quad C' = \frac{E}{r + ng}.$$

If r *is small* $C = nC'$ nearly, or the current is reduced by the high resistance galvanometer in almost the same ratio that the field is increased. There is, therefore, no gain, and the great reduction of the current renders the use of such an instrument inadvisable. If *r is large*, the resistance is not increased *n* times by the introduction of *ng* instead of *g*, and *C* is $<nC'$, or the field is strengthened in a greater ratio than the current is decreased. There is, therefore, a gain in using a high resistance galvanometer. If *r*. is very large, a galvanometer of many turns must be used to obtain any deflection at all.

45. Divided Circuits (§ 353).

It is at first difficult to understand how introducing another resistance in parallel circuit can reduce the total resistance, but it is to be recollected that the current possesses a greater number of paths to flow through. By the law that the current divides proportionally to the conductivities of the branches, $\dfrac{1}{r}$ goes through one branch and $\dfrac{1}{r'}$ through the other. Let *R* be a resistance such that the same current would pass through it as through both *r* and *r'*. Then

$$\frac{1}{R} = \frac{1}{r} + \frac{1}{r'} \text{ or } R = \frac{rr'}{r + r'} = \frac{r'}{1 + \dfrac{r}{r'}} = \frac{r}{1 + \dfrac{r'}{r}}.$$

But $\dfrac{r'}{1 + \dfrac{r}{r'}}$ is less than *r'* and $\dfrac{r}{1 + \dfrac{r'}{r}}$ is less than *r*. The resistance of a divided circuit is, therefore, always *less* than that of any of the resistances entering into it. If there are three conductors, *r*, *r'*, *r''*, we have

$$\frac{1}{R} = \frac{1}{r} + \frac{1}{r'} + \frac{1}{r''} = \text{conductivity,}$$

and

$$R = \frac{rr'r''}{rr' + rr'' + r'r''}.$$

To find the current in each branch : Let C be the total current, B the battery resistance, C' be the current in r', and C'' that in r''. Then

$$C = \frac{E}{B + \dfrac{r'r''}{r' + r''}} = C' + C''.$$

The currents are inversely as the resistances through which they flow. Taking, therefore, the resistances between A and B, Figure 129,

$$C : C' :: r' : \frac{r'\,r''}{r' + r''} \quad \text{or } C' = C\frac{r''}{r' + r''} \quad \cdot \quad \cdot \quad (1)$$

also $\quad C : C'' :: r'' : \dfrac{r'\,r''}{r' + r''} \quad \text{or } C'' = C\dfrac{r'}{r' + r''} \quad \cdot \quad \cdot \quad (2)$

46. Shunts.

These formulas are of great importance in the use of shunts for galvanometers. If a current is so powerful that there is danger of its injuring the galvanometer coils, or if it produces a deflection too near 90°, the galvanometer may be shunted by introducing a resistance in parallel circuit, so that less current will pass through the galvanometer.
Letting C be the current when the galvanometer is B unshunted, C' the total current when

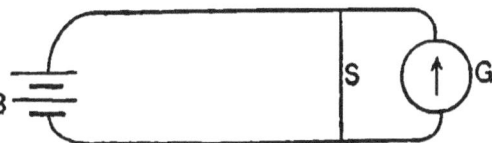

Fig. 14.

the galvanometer is shunted, C_g and C_s the currents

through the galvanometer and shunt respectively, G the resistance of the galvanometer, S that of the shunt, and R all other resistance,

$$C = \frac{E}{R + G} \qquad C' = \frac{E}{R + \dfrac{GS}{G + S}}.$$

But as $\dfrac{GS}{G + S}$ is less than G, C' is greater than C, or the introduction of the shunt has *increased* the total current in circuit. The current through the galvanometer is from (1), Note 45,

$$C_g = C' \cdot \frac{S}{G + S} = \frac{E}{R + \dfrac{GS}{G + S}} \times \frac{S}{G + S}$$

$$= \frac{ES}{R(G + S) + GS} \quad \cdot \quad \cdot \quad \cdot \quad \cdot \quad \cdot \quad (3)$$

If the deflections of the galvanometer are proportional to the currents,

$$d : d' :: C : C_g :: \frac{E}{R + G} : \frac{E}{R + \dfrac{GS}{G + S}} \times \frac{S}{G + S}$$

$$\therefore d : d' :: R(G + S) + GS : (R + G) S \quad . \quad (4)$$

These formulas may be simplified if $\dfrac{G + S}{S}$ be called

u. This proportion is sometimes called the *multiplying power* of the shunt. By its use (3), the current through the galvanometer becomes $\dfrac{C'}{u}$. The current through the shunt is $C' \dfrac{G}{uS}$. The resistance of the shunted galvanometer $\dfrac{GS}{G + S}$ becomes $\dfrac{G}{u}$, and the resistance necessary to add to the circuit to retain the same total current is $G - \dfrac{G}{u} = G\left(\dfrac{u-1}{u}\right)$. This ratio u is that between the sensibility of the shunted and the unshunted galvanometer. Thus, if the resistance of a shunt which will reduce the sensibility of a galvanometer of 1,000 ohms one hundred times is required,

$$u = 100 = \frac{1000 + S}{S} \quad \therefore S = \frac{1000}{99} = 1.01 = \frac{1}{99}G.$$

If the current is kept the same by adding a resistance of $G\left(\dfrac{u-1}{u}\right)$, $C = C'$, and $\dfrac{C}{u}$ will pass through the galvanometer.

47. Kirchhoff's Laws (§353).

The application of these useful laws may be illustrated by the figure. The following equations are derivable. From the first law, that *in any network of wires the algebraic sum of the currents meeting at a point is zero.*

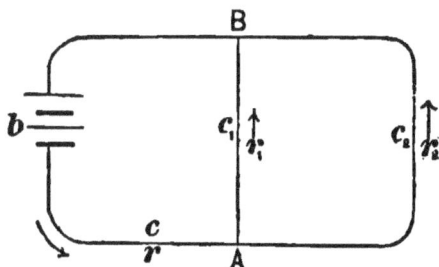

Fig. 15.

At A $\quad c = c_1 + c_2$
At B $\quad c_1 + c_2 = c.$

From the second law, that *in any closed circuit the electromotive force is equal to the sum of the separate resistances, each multiplied by the strength of current flowing through it.*

In the left hand circuit $c\,(r + b) + c_1\,r_1 = E.$
In the right hand circuit $c_1\,r_1 - c_2\,r_2 = $ zero.

48. Fall of Potential (§ 357).

Let *A* and *B* be the poles of a battery. They will have different potentials, numerically equal but of opposite signs, and the battery may be considered to preserve a constant difference of potential between them. Connect the points *A* and *B* through an external resistance. The potential will fall along this resistance, and it is required to find the potential at any point.

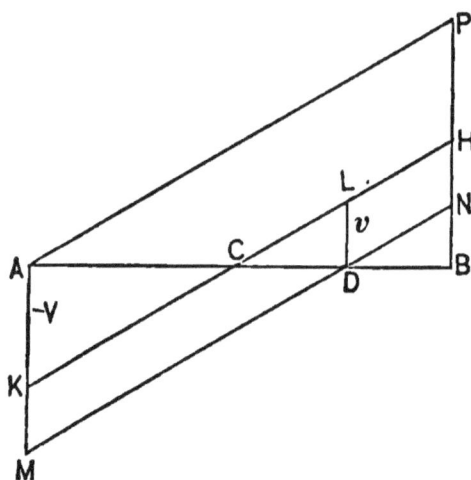

Fig. 16.

Draw *A B* to represent in length the value of the external resistance, and at *A* and *B* erect perpendiculars, one positive and the other negative, to represent proportionally the potentials at these points.

From Kirchhoff's second law, in any part of a circuit of resistance *r*, $E = Cr$. In another part of resistance r', $E' = C'r'$, *E* and *E'* representing the difference of potentials between the ends of the portions of the circuit considered. If the resistances are both in the same circuit $C = C'$, hence $E : E' :: r : r'$, or the differences of potential in the same circuit are proportional to the resistances through which they act. Assuming one potential to be zero, the potentials at the other points vary directly as the resistances separating them from the point of zero potential, and as in the figure the horizontal line represents resistance and the ordinate *BH* potential, it is evident that ordinates at other points cut off by the line CH will correctly represent the potential at these points, as they and they only will satisfy the above proportion. The line CH represents, therefore, the fall of potential in the resistance BC. If CH is prolonged to K the triangles $I^{.}AC$ and CHB are similar, and since by hypothesis AK is equal to BH, $AC = BC$, or C, the point of zero potential, is midway between the poles, and the line HK represents the fall of potential along the resistance AB. It is to be noticed that in the figure the difference ot potential between A and B is $BH + AK$; or if $-V$ is the potential at A and $+V$ at B, the difference of potential is $V - (-V) = 2V$. The difference between A and D is $v - (-V) = V + v$. If the point C is connected with the earth, as it is already at zero potential, the potential is not changed anywhere in the circuit. If, however, another point D, whose potential is $+v$, is connected to earth its potential is lowered by the amount v, and as the battery preserves a constant *difference* of potential between A and B, the absolute potential of all points in the circuit is also lowered by the amount v. The fall of potential, is, therefore, the same, and is represented by drawing a line MN through D parallel to HK, and the potential at any point is the length of the ordinate at that point intercepted by MN.

The difference of potential between A and B is now $BN + AM =$

$$(V - v) + (V + v) = 2V \text{ as before.}$$

If the negative pole A of the battery is connected to earth, the potential of all points of the circuit is raised by the amount V, and the line of potential assumes the position AP. The potential of every point of the circuit is now positive, but the differences of potential are the same as at first.

49. Wheatstone's Bridge (§ 358).

Connect the poles of a battery by two resistances, $P Q$

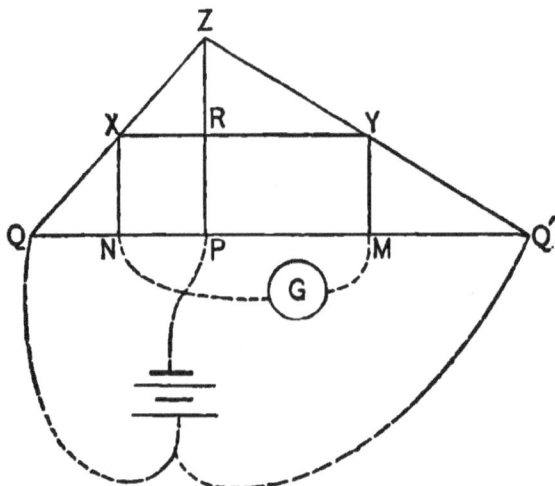

Fig. 17.

and PQ', and at P erect a perpendicular to represent the difference of potential due to the battery. Then the lines ZQ and ZQ' will represent the fall of potential in the resistances PQ and PQ'. The potential at any point on QQ' being the ordinate cut off by the lines ZQ or ZQ', if a galvanometer be joined to two points N and M at which the ordinates NX and MY are equal there will be no deflection of the needle. But from the figure, since XY is parallel to QQ', the triangles ZRX and XNQ are similar

and $$NQ : RX :: NX : RZ;$$

also $$MQ' : RY :: MY : RZ; \text{ but } NX = MY$$

$$\therefore NQ : RX :: MQ' : RY,$$

or $$A : B \ :: C : D,$$

as in Fig. 130 and § 358. When, therefore, a galvanometer joined to the junctions of two pair of resistances through which a current is flowing shows no deflection, the resistances are proportional to each other.

An important fact somewhat difficult to understand at first is apparent from Fig. 17. The resistances PNQ and PMQ (see also Fig. 130) are unequal, while having the same electromotive force acting in each. The currents in the two branches are therefore unequal. Beginners are liable to regard the balance in Wheatstone's Bridge as due to an equality of currents; but this is wrong, the equality of potentials at the galvanometer terminals being the condition of balance. This is equivalent to saying that there is no current through the galvanometer, offering another method of proof as follows:

50. Proof of Theory of Wheatstone's Bridge by Kirchhoff's Laws.

Let the currents in the different branches be represented by

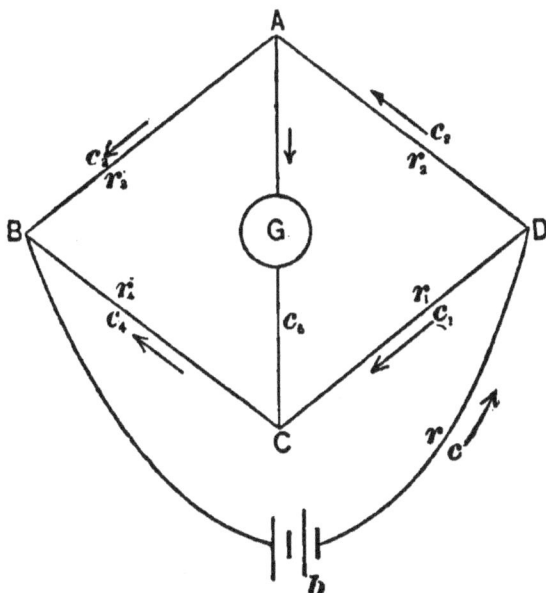

Fig. 18.

c, c_1, c_2, etc., and the corresponding resistances by r, r_1, r_2, etc. Let E be the $E.\ M.\ F.$ and b the resistance of the battery. From the first law

$$
\begin{aligned}
\text{At } A \quad & c_2 = c_3 + c_5 & & \dots\dots\dots\dots & (1)\\
\text{`` } B \quad & c = c_3 + c_4 & & \dots\dots\dots\dots & (2)\\
\text{`` } C \quad & c_4 = c_1 + c_5 & & \dots\dots\dots\dots & (3)\\
\text{`` } D \quad & c = c_2 + c_1 & & \dots\dots\dots\dots & (4)
\end{aligned}
$$

By the $\Big\{$ In ABC, $\ c_3\,r_3 - c_4\,r_4 - c_5\,G = 0\quad\dots\dots\ (5)$
second $\Big\{$ `` ADC, $\ c_2\,r_2 + c_5\,G - c_1\,r_1 = 0\quad\dots\dots\ (6)$
law, $\Big($ `` $BbDC$, $\ c\,(b + r), + c_1\,r_1 + c_4\,r_4 - E = \ 0.\ (7)$

Adjust r_1, r_2, r_3 and r_4 until the galvanometer shows no deflection, then c_5 is zero. The above equations being general, substitute zero for c_5 and they become

(1) $c_2 = c_3$	(5) $c_3\,r_3 = c_4\,r_4$ or $c_2\,r_3 = c_4\,r_4$. (8)
(3) $c_4 = c_1$	(6) $c_2\,r_2 = c_1\,r_1$ or $c_3\,r_2 = c_4\,r_1$. (9)

Dividing (8) by (9) $\dfrac{r_3}{r_2} = \dfrac{r_4}{r_1}$.

When, therefore, the galvanometer shows no deflection, the arms of the bridge are proportional. The accuracy of the measurement depends, of course, on the sensibility of the galvanometer.

51. Measurement of Electromotive-Force (§ 360).*

(a) WHEATSTONE'S METHOD.

Let G be the resistance of the galvanometer, R the remaining resistance in circuit, and ρ the resistance necessary to add to the second battery to reduce its deflection to that of the first. Then for a deflection d

---

* The memoranda of Notes 51, 52 and 53 are confined strictly to an explanation of the methods of measurements referred to in " Thompson's Elementary Lessons," § 360, 361 and 362 Other methods may be obtained from any work on electrical measurements.

$$\frac{E}{R + G} = \frac{E'}{R + \rho + G}$$

whence $\quad \dfrac{R + G}{E} = \dfrac{R + \rho + G}{E'} \quad$ . . . . . (1)

To bring the deflection to d', extra resistances have to be added. Represent these by r and r'. For the deflection d'

$$\frac{E}{R + G + r} = \frac{E'}{R + \rho + G + r'},$$

whence $\quad \dfrac{r}{E} + \dfrac{R + G}{E} = \dfrac{r'}{E'} + \dfrac{R + \rho + G}{E'} \quad$ . . . (2)

Substituting (1) in (2) it reduces to

$$\frac{r}{E} = \frac{r'}{E'}; \text{ or, } r' : r : : E' : E.$$

(b) CLARK'S METHOD.

Clark's method requires three cells. E furnishes a current and has the highest
E. M. F., E' is a
standard cell, gener-
ally that of Clark
(§ 177), and E'' is the
cell to be tested. If
E'' has a higher
E. M. F. than 1.457
two or more Clark's
cells must be used
at E'. The similar
poles of the cells
a r e connected to
A. In measuring,

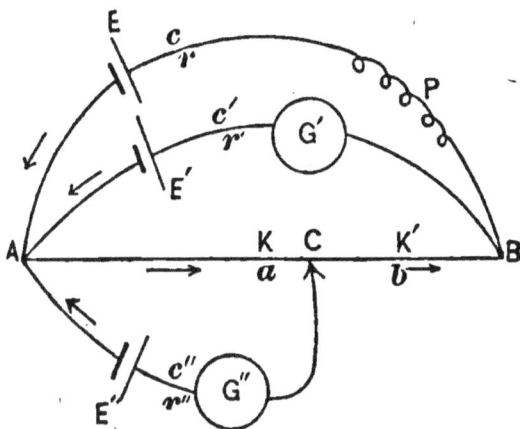

Fig. 19.

E'' is first disconnected and the needle of G' is brought to zero by adjusting P. E'' is then connected at A and the sliding piece C is moved along the resistance A B, shifting contact until G'' also shows no deflection.

From Kirchhoff's laws :

$$\text{At } A \qquad c + c' + c'' - K = o \quad \ldots \quad \text{(1)}$$
$$\text{`` } C \qquad K - c'' - K' \quad = o \quad \ldots \quad \text{(2)}$$
$$\text{`` } B \qquad K' - c' - c \quad = o \quad \ldots \quad \text{(3)}$$
$$\text{In circuit } G'AB \quad c' \, (r' + G') + Ka + K'b - E' = o \quad \text{(4)}$$
$$\text{`` } \quad A\,G''C \qquad c'' \,(r'' + G'') + Ka - E'' = o \quad \text{(5)}$$

But when adjustment is secured c' and c'' are each zero. Substituting these values

from (1) $c = K$ from (2) $K = K'$ from (3) $c = K'$
from (4) $Ka + K'b = E'$; or, $K\,(a + b) = E'$ $\ldots$ (7)
from (5) $Ka = E''$ $\ldots\ldots\ldots\ldots\ldots$ (8)

Combining (7) and (8)

$$E' : E'' : : a + b : a.$$

(c) QUADRANT ELECTROMETER.

The two poles of a standard cell are connected to the quadrants, the same pole being in connection with opposite segments, as 1 and 3, Fig. 101. The deflection of the needle is then noted. The standard cell is then disconnected and the one to be tested substituted in the same way. From the ratio of the deflections, the ratio of the electro-motive forces may be obtained. Care must be taken that the needle is electrified to the same potential in the two measurements.

52. Measurements of Internal Resistance (§ 361).

(a) Connect the battery in circuit with a galvanometer and a box of resistance coils, the resistances being B, G and R respectively. Note the deflection d in the galvanometer. Increase R to R' and note the deflection d'. Then if a tangent galvanometer is used

$$\tan d : \tan d' : : \frac{E}{B + G + R} : \frac{E}{B + G + R'} \cdot$$

Reducing, $B = \dfrac{R' \tan d' - R \tan d}{\tan d - \tan d'} - G.$

If G is of no resistance, and the first deflection was taken with no other appreciable resistance than that of the battery in circuit

$$B = \frac{R' \tan d'}{\tan d - \tan d'} \cdot$$

(c) MANCE'S METHOD.

The cell whose resistance is to be measured is placed in the bridge as an unknown resistance, and a galvanometer and key (not in the same branch) are connected as in the figure. The arrows denoting the direction of the current, it is at once evident that a current passes through the galvanometer all the time, and continues to do so unless the resistance in c and d becomes zero. Every change of resistance in a, c or d will affect the current flowing through G, as will also the opening or closing of the key k when any difference of

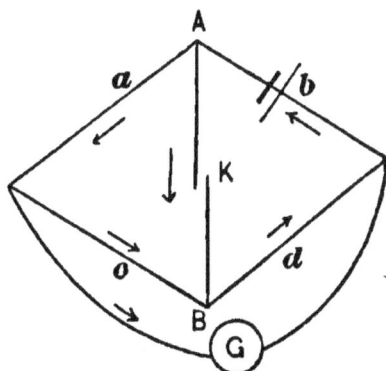

Fig. 20.

potential exists between A and B. If on pressing k there is no *change* in the galvanometer deflection, it follows that A and B are at the same potential and that consequently

$$a : b : : c : d \text{ or } b = \frac{ad}{c},$$

the common relation of the Wheatstone bridge. If the galvanometer is a sensitive one, the deflection will be nearly $90°$ and as in that position a change of current produces but little effect on the needle it is necessary to reduce the deflection either by the use of a magnet or by shunting the galvanometer.

53. Measurement of the Capacity of a Condenser (§ 362).

(a) Let $V =$ the known potential,
$\quad\quad x =$ the unknown capacity,
$\quad\quad C =$ the capacity of the standard condenser.

The original charge in the condenser of unknown capacity is Vx. On connecting the standard condenser the capacity is increased, and the quantity being the same the potential falls to V'

$$\therefore Vx = V' (x + C),$$
$$\text{and } x : x + C :: V' : V.$$

(b) The impulse acting to deflect the needle varies as the quantity of electricity passing. If the two condensers are charged from the same cell the potential is equal. Consequently (Note 5),

$$\sin \tfrac{1}{2} d : \sin \tfrac{1}{2} d' :: VC : VC'$$
$$:: C : C'.$$

(c) There is apparently some error in stating this method in "Thompson." It probably refers to a common and quite accurate method applied by Sir Wm. Thomson to the measurement of the capacity of cables.

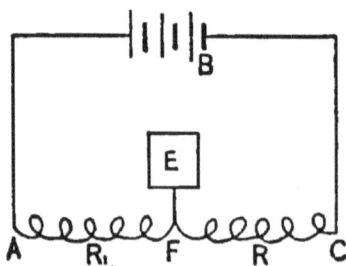

Fig. 21.

If the poles of a battery are connected through a high resistance $R_1 + R$ and a point F is connected with earth, its potential becomes zero and points on either side separated from it by equal resistances are at equal but opposite potentials. If the resistances are unequal,

Potential at A : Potential at C : : R_1 : R,

or $\quad V_1 : V_2 :: R_1 : R$,

$$\therefore V_2 = \frac{V_1 R}{R_1}.$$

Charge the two condensers by making contact on opposite sides of F at such points that their charges just neutralize. If A and C are the points,

$$V_1 C_1 = V_2 C_2$$

$$= \frac{V_1 R_2}{R_1} C_2 \quad \therefore C_2 = \frac{R_1}{R} C_1.$$

(d) Charge the condenser and discharge it through a galvanometer, noting the deflection; charge it again to the original potential and allow it to discharge slowly through a very high resistance. After discharging a definite time T, note the deflection it will give when connected directly to the galvanometer. Then

$$\text{Capacity} = \frac{T}{2.303 \; R_1 \log \dfrac{V}{v}} \; ;$$

V and v are potentials, but as the ratio only is needed, this can be obtained from the ratio of the two deflections.

54. Determination of the Ohm (§ 364).

If a coil is rotated in a magnetic field a current is induced, which deflects a needle at the centre of the coil. The force exerted by this current may be shown to be $\dfrac{L^2 VH}{4K^2 R}$, and the moment acting on a magnet deflected through an angle θ is $\dfrac{L^2 VH}{4K^2 R} \, ml \, \cos \theta$, where L is the length of the coil, V the velocity of rotation, H the intensity of the field, K the radius of the coil and R its resistance. The moment of the force of the earth's magnetism tending to bring the deflected magnet into the meridian is $Hml \sin \theta$. Equating these moments when the needle maintains a steady deflection,

$$\frac{L^2 VH}{4K^2 R}\, ml \cos \theta = Hml \sin \theta;$$

$$\therefore R = \frac{L^2 V}{4K^2 \tan \theta}.$$

The value of R is given in absolute units of resistance.

55. Practical Electromagnetic Units of Heat (§ 367) and of Work (§ 378).

As shown in Note 7, the measure of the work done in moving a quantity of electricity is the product of the quantity of electricity by the difference of potential through which it is moved, or

$$\text{Work} = \text{Quantity}\ (V_1 - V_2) = Ct \times E.$$

If the current, time, and electromotive force are all expressed in absolute units, the work is given in ergs. Substituting for the practical units of current and electromotive force, the ampère and the volt, their values in absolute units,

$$\text{Work} = 10^{-1} \times 10^8 = 10^7 \text{ ergs per second.}$$

This equation gives a practical unit of work, corresponding to the practical units of current and electromotive force referred to, and Dr. Siemens has proposed that it be called the "*watt*." This suggestion has been well received, and the unit is coming into use in electrical calculations. The value of the watt is 10^7 ergs, and it may be defined as the work done by a current of one ampère in a portion of the circuit in which the potential falls one volt.

1 H. P. English $= 33000$ foot-pounds per minute.
$= 550$ "   "   " second.
$= 76.04$ kilogrammetres per second.
$= 76.04 \times 10^6$ gramme-centimetres per second.
$= 76.04 \times 981 \times 10^6$ ergs per second.
$= 746 \times 10^7$ ergs $= 746$ watts.

To find, therefore, the work done by an electric current in any portion of a circuit, *measure the difference of potential in volts between the ends of the portion considered, multiply it by the current in ampères, and divide by* 746. The quotient is the work in horsepower. In the above calculation the dimensional equation of work cannot be used for the change of foot-pounds to kilogrammetres, as these are statical units, and the dimensions are for dynamical units only.

As heat and work are the same, the heat measured in ergs given off in any part of a circuit is also CEt, or substituting the value of E from Ohm's Law,

Heat in ergs $= C^2Rt$.

This formula is more generally used, as it gives the amount of heat developed in a resistance. Substituting as before the values of the ampère and ohm in absolute units, the heat is measured in practical units, each of which is of the value of 10^7 ergs. This unit is sometimes called the "*joule*," and may be defined as the heat evolved in one second by a current of one ampère in a resistance of one ohm.

To obtain the value of the joule in terms of the more

common thermal unit, which is the amount of heat necessary to raise one gramme of water one degree centigrade, it is necessary to use the mechanical equivalent of
heat, which is 42,400 gramme-centimetres. A gramme of
water in falling 42,400 centimetres acquires sufficient
energy to raise its temperature one degree centigrade if
suddenly stopped. The thermal unit, therefore, equals
42,400 gramme-centimetres or 42,400 × 981 ergs = 4.16
× 10^7 ergs, or approximately 4.2 × 10^7 ergs. But as
the joule is 10^7 ergs, the water-gramme-degree heat unit is
equal to 4.2 joules, or the joule is .238 of the thermal
unit referred to.

The watt and joule are of the same value, but one ex-
presses the energy given off in a circuit in terms of power,
and the other in heat.

V. ELECTRIC LIGHTS.

56. The Voltaic Arc (§ 371).

If a powerful electric current is broken at any point, there is a bright spark at the break, and if the two terminals of the circuit on each side of the break are kept at a constant short distance, a steady light or arc will be produced. The color of the light varies with the materials between which the arc is formed. The heat produced is the highest known, and most substances fuse so readily in the arc that they cannot be used for electrodes. For this reason gas carbon, which is practically infusible and of comparatively low electrical resistance, is universally used for the pencils of arc lights. If the image of the arc is thrown on a screen the greater part of the light is seen to be due to the carbon points being heated white hot, the arc itself being generally bluish and less brilliant. If a magnet is brought near the arc, the interaction of magnets and currents is well illustrated by the movements of the arc to one side, and it is even possible to deflect it until it assumes the position of a blowpipe flame.

The electric light was prevented from coming into everyday use, so long as the current it required could not be obtained from any cheaper source than the voltaic battery. The invention of dynamo machines, by greatly decreasing the cost of electric energy, rendered the extensive use of the arc light a possibility. This may be best illustrated by an example.

A Number 7 Brush machine works sixteen arc lights in

series. The *E. M. F.* of the circuit is 839 volts, the internal resistance of the machine 10.5 ohms, the resistance of the lights and leading wires 73 ohms and the current 10 ampères. Assuming the electromotive force of a Grove cell to be 1.8 volts and its internal resistance $\frac{2}{10}$, the number of cells necessary to do the work of the machine may be calculated.

To obtain the *E. M. F.* the number of cells necessary is $\frac{839}{1.8} = 466$ in series. But 466 cells in series have an internal resistance of 93.2 ohms, and that the current may be the same, enough cells must be introduced in arc to reduce the battery resistance to that of the machine. $\frac{93.2}{10.5}$ is nearly nine, and it is therefore necessary to use 466 groups in series, each group containing 9 cells in arc or 4,194 cells in all. This number of cells would cost ten times as much as the machine, and could not keep up the current for more than two or three hours.

Calculating in the same way with a constant battery, assuming the *E. M. F.* of a gravity cell to be 1.08 volts and its internal resistance 5 ohms, it will be found necessary to arrange 777 groups in series, each group containing 369 cells in parallel arc, a total of 286,713 cells, a greater number in all probability than are in use in the United States. The invention of the dynamo machine has therefore not only operated to diminish the cost of the electric light, but to bring it within the bounds of commercial practicability.

As a general rule, the light given out by an arc varies as the current. A small current will give no light at all, and, as stated in § 371, a certain electromotive force and current are necessary for the production of a satisfactory light. After that point a stronger current causes more light.

57. Arc Lamps (§ 372).

Gas carbon pencils are now used almost exclusively, and are frequently coated with a thin film of copper to prevent the oxidation and waste of the carbon before it becomes incandescent. Two general methods are in use for regulating the distance of the carbons from each other, one using clock-work and the other regulating directly by the current. The most widely used of the latter class is the Brush lamp, a plan of which is given in the figure. The current entering at A, divides at B into two branches which pass around the bobbin C in opposite directions, one branch being a coarse wire of low resistance and in the same circuit as the carbons, and the other branch S S being a shunt of high resistance to the lamp, connecting the terminals B and G. Inside the bobbin is a soft iron armature F, which is attached to the upper carbon. When a current passes the two branch circuits on the bobbin C tend to magnetize it in opposite directions, but the resistances and number of turns in the two circuits are so proportioned that the magnetic field due to the low resistance branch is the stronger, and the armature F is therefore attracted, lifting the upper carbon and establishing the arc. Should the carbons become too widely separated the resistance of the arc and consequently of the coarse wire circuit on C increases, diminishing the current in C and increasing that in the shunt S.

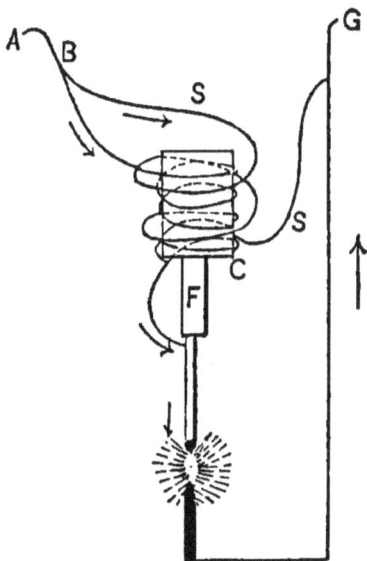

Fig. 22.

The field due to the shunt is therefore strengthened and that due to the coarse wire diminished, allowing the armature F to fall slightly, bringing the carbons nearer together. By the device of the two opposing fields, due to the coils on C being wound in opposite directions, the feeding of the lamp is therefore done automatically, and the actual distance of the two carbons varies but little. In the lamp as constructed, two bobbins are used in parallel arc, and the armature F clutches the upper carbon. A plunger moving in glycerine is also attached to the upper carbon to render the movements less sudden, and the shunt circuit S S passes around another bobbin, which, by attracting an armature, closes the main circuit, and short circuits the lamp in case the carbons are broken or the adjustment does not work. The figure is designed merely to illustrate the general method.

The Foucault regulator shown in Fig. 138 is more complicated than the Brush. The two carbons are clamped to rods which are moved by clock-work, the gearing being such that the positive carbon moves twice as rapidly as the negative. In this way the arc is kept approximately in the same position, admitting of focussing in a projecting apparatus. The clock-work consists mainly of two barrels driven by springs and acting through gearing on the rods carrying the carbons, one barrel separating the carbons and the other bringing them nearer together. The electromagnet seen in the base of Fig. 138 is in the main circuit, and its armature works a system of levers, the last one of which acts by a pawl on two small flies, each connected through the gearing with one of the barrels. If, then, the current becomes too strong, the armature is attracted, one of the flies is released and the corresponding barrel sets the clock-work in operation, separating the carbons slightly. As they approach their normal distance apart the current diminishes, and the armature moves away slightly, pawling the fly. As the carbons burn away, the current is diminished still more, the armature is less attracted, the other fly is released and the clock-work moves to bring the carbons nearer together. The arc may be formed at any height by moving the upper carbon by means of the rod at the top of the regulator.

58. Incandescent Lamps (§ 374).

That electric lighting may be of universal utility, lamps are necessary which give only a moderate light. Arc lights cannot be made to work with certainty and economy at low illuminating power, and the incandescent lamp is therefore coming into use. In this case the light is given off from a portion of the circuit which is heated white hot by the passage of the current, and this requires that that portion shall be of high resistance (§ 367) and practically infusible. After long experimenting, carbon has been fixed upon as the best material, and it is now used in all types of incandescent lamps. A small fibre of carbon is obtained by heating some vegetable substance out of contact with the air and driving off all volatile matter. Edison uses bamboo fibre, Swan cotton thread, Lane-Fox a grass fibre, and Maxim paper. This fibre is mounted in a vacuum, on the perfection of which depends greatly the time the lamp will last, the presence of a small amount of oxygen insuring the destruction of the fibre by chemical action. Even in a perfect vacuum the fibre will eventually give way on account of what is known as the " Crookes's effect," a transfer of molecules of carbon across from one heel of the carbon to the other. Alternate currents by wearing each heel away equally tend to lengthen the lifetime of the lamp, but in every case the final giving way of the fibre is a matter of time only. The lifetime may be greatly prolonged by working the lamp below its normal power. The life of an Edison lamp, working with its normal current, is now (1883) probably about 1,000 hours.

As pointed out, the condition that much heat should be developed in any part of a circuit is, that that part should be of high resistance. If several lamps were placed in series, the resistance of the circuit would be so great that the current would be insufficient, and they are, therefore, placed generally in parallel arc, reducing the external resistance to such an extent that each lamp has it full share of current. Assume a battery or machine giving an electromotive force of 200 volts at its terminals and having an internal resistance of five ohms. If two lamps each of 100 ohms resistance, and requiring one ampère to give their normal light, were placed in the exterior circuit in series, the current would be $\frac{200}{200 + 5} = .975$ ampères—not sufficient to work the lamps at their normal standard. By placing the lamps, however, in parallel arc the same machine would, under the same conditions, work twenty lamps, thus, $\dfrac{200}{\frac{100}{20} + 5} = 20$ ampères, or one in each lamp.

That all the lamps may give the same light, the resistances must be equal, otherwise some will have stronger currents passing through them than others. Placing a larger number in parallel arc will still further reduce the resistance, strengthening the current, but increasing the number of parts into which the current must divide. Thus in the case above, the machine cannot work 21 lamps, for $\dfrac{200}{\frac{100}{21} + 5} = 20.47$ ampères, or only .975 to each lamp.

A most important feature in incandescent lighting is the change of resistance of carbon by temperature. The resistance of metals increases with the current passed through them, while that of carbon decreases, the reduc-

tion in an incandescent lamp being between 40 and 50
per cent. As the good working of any system of lighting
depends on a correct adjustment of resistances, this peculiarity of carbon must be allowed for. The following
measurements made on an Edison lamp at the U. S. Tor-
pedo Station, Newport, illustrate the change of light and
of resistance accompanying a change of current.

Current.	Resistance.	Candles.
.000	134.	Lamp cold.
.114	110.5	Cherry red.
.203	94.7	Bright red.
.309	86.7	Orange.
.440	81.8	1.8 Candles.
.680	73 5	7.2 "
.750	72.8	11.4 "
.810	70.6	16.4 "
.890	69.9	21.0 "

It is seen that a current of $\frac{3}{10}$ of an ampère produced a
light just perceptible, but that every increase after that
produced a much greater proportionate increase of light.
The lamps are, therefore, more economical of energy the
more light they give, but working with high currents in-
sures the rapid destruction of the lamp. At present the
Edison lamps are the most economical, yielding under
good working conditions about twelve sixteen-candle
lamps per horse power of current (See § 378).

6

VI. ELECTRO-MAGNETIC INDUCTION.

59. Induction Currents Produced by Currents (§ 393).

Take two coils of insulated wire, wound so that one may be inserted inside the other, and place a battery in circuit with one and a galvanometer with the other. There is no communication between the two coils, the wire being thoroughly insulated, but if the coil in circuit with the battery is slowly inserted in that in circuit with the galvanometer (Fig. 147), the latter will show a deflection, which is due to what is known as an *induced* current. Experiment shows that this current is not continuous, but exists only when one coil is moved near the other. If the small coil is inside the larger, no current is observed in the latter so long as the current in the former is constant, but if it is broken, an induced current is observed in the outer or " *secondary* " coil, flowing in the *same* direction as that formerly existing in the " *primary* " or inner coil. If the circuit is closed in the latter while it is inside the secondary coil, the induced current in the secondary is in the *opposite* direction to that in the primary. If the primary coil is gradually removed, a current is induced in the secondary in the *same* direction as that in the primary. There is no current in the secondary, unless there is some *change* either in the strength of the primary current, or in its position relatively to the secondary coil.

The same effects are produced by moving a magnet in or near a coil in circuit with a galvanometer. If in this case the direction of the Ampèrian currents (see Note 42) be considered, the induced current in the coil will be in the *opposite* or in the *same* direction as that of the Ampèrian currents, as the magnet is introduced or withdrawn from the coil. (See Fig. 146.)

Prof. Thompson generally uses the word "*direct*," as applied to currents, to represent a positive current, one flowing in the direction in which the hands of a watch move, but in explaining these experiments he has used it in another sense as meaning a current in the same direction, and confusion is apt to result. Whether a *direct* or *inverse* current exists in a secondary circuit depends not only on the motion of the primary coil, but on the direction of the current in it, but his explanation does not include the latter at all. Reserving the terms *direct* and *inverse* to apply to the *positive* and *negative* directions, the summing up on Page 360 may be expressed thus :

By means of a	Momentary currents in the opposite direction are induced in the secondary circuit	Momentary currents in the same direction are induced in the secondary circuit
Magnet	while *approaching*.	while *receding*.
Current	while *approaching*, or *beginning*, or *increasing*.	while *receding*, or *ending*, or *decreasing*.

These rules, when applied to magnets, call for a conception of imaginary currents flowing around them. By considering the induction of the current to be due to the movement with or against magnetic forces, all the above relations may be expressed by the rule given in ¿ 394 (i.).

"*A decrease in the number of lines of force which pass through a circuit produces a current round the circuit in the positive direction (i. e., a 'direct' current); while an increase in the number of lines of force which pass through the circuit produces a current in the negative direction round the circuit.*" In the application of this rule, care must be taken to look along the lines of force in their positive direction, that in which a north pole tends to move, as a current which appears to be direct when viewing it from one side is inverse if seen from the other, and the rule is worthless if misapplied.

60. Determination of the Induced Electromotive Force (§ 394, ii.).

From the foregoing rules, it is seen that a current is induced in a closed circuit only when there is some change in the number of magnetic lines of force enclosed by the circuit. This change may be due either to the motion of the pole, to that of the circuit, or to the change of current strength, if the field is due to a current. Each of these three cases requires the expenditure of energy in some form, and without such expenditure there is no induced current. We are therefore led to look on the energy of the induced current as directly derivable from the energy expended in producing the change in the field.

Let a current flow from a battery through a coil placed in a magnetic field. As already shown (Note 36) the work done in producing any displacement of the coil is $- C (N_2 - N_1)$, or if dN is the change in the number of lines of force passing through the circuit in the time dt, this may be written as $- C. dN$. The current in flowing through the coil heats it, the amount of heat being (§ 36/. $C^2 R . dt$. If the coil is placed so that the lines of force of the field pass through it in the wrong direction, it will move, doing work which is due to the original energy of the current in the circuit. Hence

Energy of the current = heating effect + work done ;

or, $CE \cdot dt = C^2 R \cdot dt + C \cdot dN$

$$\therefore E = CR + \frac{dN}{dt} \quad . \quad . \quad . \quad . \quad . \quad (1)$$

CR is the E. M. F. in the coil, while E is that originally due to the cell. The former is less than the latter by $\frac{dN}{dt}$, which must also be an E. M. F. and due to the work done by the coil. As a result, therefore, of the motion of the coil in the field, the E. M. F. originally in circuit is diminished by the amount $\frac{dN}{dt}$. The E. M. F. remaining in the circuit is

$$CR = E - \frac{dN}{dt} \quad . \quad . \quad . \quad . \quad . \quad (2)$$

The induced E. M. F. is therefore measured by the *rate* of change in the number of lines of force which pass through the circuit, and is *opposite in direction* to that originally existing, which caused the motion. By increasing this rate, by diminishing dt, or, what is the same thing, making the velocity of motion greater, $\frac{dN}{dt}$ may be made to equal or exceed E, and the direction of the induced current would therefore be the same whether there was any E. M. F. to be overcome or not. If a battery current flows, the induced current diminishes it ; if there is no battery current, the coil would have to be moved by external agency, and the induced current is in the opposite direction to a current which would cause the motion. This is seen directly from (2). If $E = O$

$$CR = -\frac{dN}{dt} \quad \ldots \ldots \ldots \quad (3)$$

The fact that the induced current acts in an opposite direction to that causing the motion is easily deduced from the principle of conservation of energy, for, if a current flowing in a given direction caused motion of the circuit, and this motion induced a current in the same direction as the original, it would increase the motion and consequently the energy of the system. Lenz's law, given in § 396, is therefore a direct result of the conservation of energy.

On the supposition that the original E. M. F. is zero, or that there is no current flowing when the coil is at rest,

$$O = C^2R.dt + C.dN. \quad \ldots \ldots \quad (4)$$

The original energy being zero, the energy of the current when the coil is moved can be obtained only from the work done in moving the coil, or

Work done in moving coil = heating effect + work done by the coil.

If the coil does no work, the total energy appears as heat in the circuit.

The E. M. F. induced in the circuit being $\frac{dN}{dt}$ the current is equal to $\frac{dN}{R.dt}$, in which R is the total resistance in circuit.

61. Practical Rule for Direction of Induced Current (§ 395).

From the rule given in § 186, "Suppose a man swimming in the wire with the current, and that he turns so as to look along the lines of force in their positive direction, then he and the conducting wire with him will be urged towards his left," combined with Lenz's Law (Note 62), the following rule for the direction of the induced current in a conductor is easily deduced. *Suppose a man swimming in any conductor to turn so as to look along the lines of force in their positive direction; then if he and the conductor be moved toward his left hand he will be swimming against the current induced by this motion; if he be moved toward his right hand the current will be with him.* Through some error this rule is given incorrectly in § 395, and differs there from the rule as given by Prof. Thompson in his Cantor lectures.

62. Lenz's Law (§ 396).

In Note 60 it was shown that in accordance with the principle of the conservation of energy, the induced current resulting from any motion of a conductor must be in the opposite direction to that of the current which would cause the motion. Lenz deduced this relation independently, and his statement that "*in all cases of electromagnetic induction the induced currents have such a direction that their reaction tends to stop the motion which produces them*" is known as Lenz's Law. As an illustration of the use of the law, suppose a magnet to be inserted in a hollow coil. The induced currents must be in such a direction as to oppose the motion. As opposite currents repel each other, the current induced in the coil will be opposite in direction to the Ampèrian current of the magnet. If the magnet is withdrawn, the withdrawal would be opposed by a current in the same direction as the Ampèrian current, and by the law a current would therefore be induced in that direction. The same reasoning applies to currents.

63. Mutual Induction of Two Circuits (§ 397).

In § 320 and Note 39 it was seen that two circuits tended to place themselves in such a position as to inclose as many of each other's lines of force as possible, and the number inclosed when each carried unit current was denoted by M. It has since been shown that any movement of either circuit induces a current in the other, and consequently changes the value of M. This quantity is therefore called the " coefficient of mutual induction." By a course of reasoning similar to that in Note 18 it may be shown that the force just outside the plane of a voltaic circuit of unit area is $4\pi C$, and if S is the area inclosed by the circuit the total force is $4\pi CS$, which becomes $4\pi S$ when C is of unit strength and $4\pi nS$ when there are n turns each of area S. But as the number of lines of force is numerically equal to the strength of the field, this number would therefore be $4\pi S$, and if *all* these lines passed through the other circuit, the maximum value of M is $4\pi S$ when the two circuits are coincident.

64. Self-Induction (§ 404).

The *extra current* is a current induced in the same conductor in which the original current flows. By Lenz's law, or by the table in Note 59, when a current begins in a conductor, a momentary current is induced in the opposite direction, and this phenomenon is noticeable as well in the original circuit as in another near it. The current in beginning is, therefore, opposed by an induced current in the opposite direction, and its increase is made more gradual, and more time is necessary for it to gain its full strength. The fact that the primary current is greatly reduced by the induced current accounts for the fact that when a circuit is closed, but little of a spark is seen. After

the current attains its normal strength it remains un-
affected by induction, unless acted upon by external causes ;
but if the circuit is broken, the law and table already referred to indicate a current in the same direction as the
primary, retarding its decrease. As this current practically reinforces the primary, the spark at breaking the
circuit is much brighter than at making.

The extra current being strictly an induced current is
subject to the general laws of induction. The induced E.
M. F. is therefore $-\dfrac{dN}{dt}$. If dN is constant, as it is with
any given current, the E. M. F. of the extra current varies
inversely as dt, or the more quickly the circuit is closed
or broken the greater the extra current. The value of L
in § 404 determines the other important relation, that the
self-induction varies as the *square* of the number of turns,
and as the resistance of the coil increases only as the
number of turns, the extra current is much greater the
more turns the coil possesses. This is of great importance
in enabling extra currents of high E. M. F. to be obtained
when wished, or avoided when they would be detrimental.

65. Helmholtz's Equations (§ 405).

The current is prevented by its self-induction from obtaining
its full strength immediately. The electromotive force of the
induced current is $-L \cdot \dfrac{dC}{dt}$, L being the coefficient of self-
induction, and the current acting in opposition to the primary
current is $\dfrac{L}{R} \cdot \dfrac{dC}{dt}$. In any interval of time, dt, after the circuit is
closed, the current has a strength of

$$C = \frac{E}{R} - \frac{L}{R} \cdot \frac{dC}{dt}, \text{ or } \frac{dC}{\dfrac{E}{R} - C} = \frac{R}{L} \cdot dt;$$

and the current at a time t from the instant of closing the circuit is

$$\int_0^c \frac{dC}{\frac{E}{R} - C} = \int_0^L \frac{R}{L} \cdot dt.$$

Integrating,

$$- \log \left(\frac{\frac{E}{R} - C}{\frac{E}{R}} \right)_0^c = \frac{Rt}{L} \therefore C = \frac{E}{R} \left(1 - \varepsilon^{-\frac{R}{L}t} \right).$$

66. Induction Coil (§ 398).

It is important to remember that in the induction coil there are two circuits, not only independent of each other but carefully insulated. The primary coil, or the one immediately surrounding the iron core, is of comparatively few turns, and of low resistance, that a given E. M. F. may cause as powerful a current as possible and consequently induce as many lines of force as possible through the core. If the primary current is made and broken rapidly, this number of lines is alternately added and subtracted at very short intervals, and the E. M. F. induced in the secondary coil, through the axis of which all the lines of the primary pass, is therefore very great. From the general formula $E = - \dfrac{dN}{dt}$, the induced E. M. F. in the secondary is evidently increased by using greater battery power in the primary or by making and breaking the primary circuit with greater rapidity. The E. M. F. of the secondary circuit becomes so great that extreme care has to be taken with the insulation, and parts of the coil at widely different potentials must not be brought near together.

It is noticeable that the *quantity* of electricity passing

in the secondary coil is extremely small. This is at once apparent when it is considered that the energy of the secondary coil, which may be expressed as $C'E'$, is all derivable from that of the primary coil CE and cannot exceed it. If the energy in the two coils is assumed equal,

$$C : C' :: E' : E,$$

and the enormous increase of E. M. F. in the secondary is, therefore, attended with a great reduction in the quantity of electricity passing. As all the energy of the primary current cannot be transferred to the secondary, this proportion is not strictly correct, but it illustrates the important point that an induction coil which might kill a man could not heat a wire red hot, or perform other work where *quantity* was necessary.

The condenser is made of sheets of tin foil insulated from each other by paper soaked in paraffine. Alternate sheets are connected throughout, so as to form two large coatings. The action may be understood from the figure. The current passes from the battery through W to I the

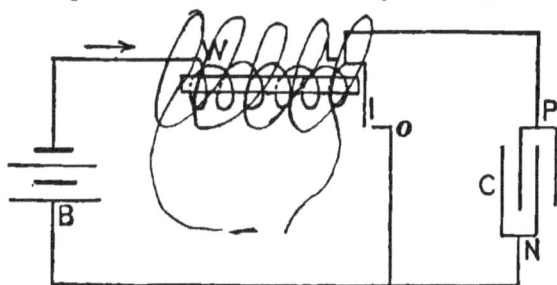

Fig. 23.

interrupter, and thence through o back to the battery. The core, on being magnetized, attracts I, breaking the current at o. If there were no condenser the extra current would leap across from I to o in a bright spark, but when the condenser is used it darts into it, charging P

positively and N negatively, but immediately afterwards the two charges re-combine, the positive charge passing from P to $L\,WBN$, demagnetizing the core and making the "break" more rapid, and also opposing the current at the "make." In this way the time dt of the break is diminished while that of the make is increased, and the E. M. F. induced in the secondary coil at the former is, therefore, much the greater. By separating the poles of the secondary circuit, they may be placed so that the break or similar current can strike across them while the make or reverse current cannot.

VII. DYNAMO MACHINES.

67. General Principle of Dynamo Machines (§ 407).

The principle underlying all production of electricity by machines, is that of Note 59, that if a coil of wire is moved in a magnetic field a current is induced in the coil. The successive machines have simply been developments of this fact, improvements having been made either in the distribution of the lines of force in the field, or in the construction and movement of the coils. In the first machines, those of Pixii, Saxton and Clarke, permanent steel magnets were used, but only a portion of the lines due to these poles were cut by the coils, and the machines were, therefore, inefficient. As a rule, the coils moved in front of the poles, intercepting the lines passing off from the poles in one direction only. The same general principle was followed in the Holmes and Alliance machines, there being a greater number of magnets and coils, with a poor disposition of the different parts. The introduction of Siemens' armature

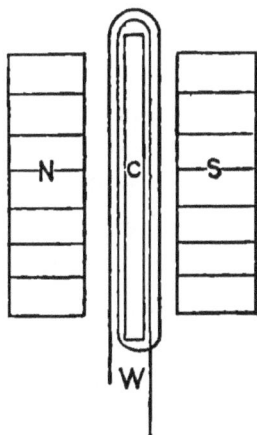

Fig. 24 A.

in 1857, was a great step in advance. *N* and *S* are the poles of several horseshoe magnets bolted together side by side, and *between* the opposite poles rotates a soft iron cylinder *C* on which the wire is coiled. This armature is thus placed in the strongest part of the field, the greater number of lines of force passing directly from N to S through the core C, and the construction admits of the coils approaching the poles very closely.   Very strong currents were obtained from machines in which this armature was used, but great difficulty was experienced from the heating of the armature.   If a disk of any metal is rotated rapidly

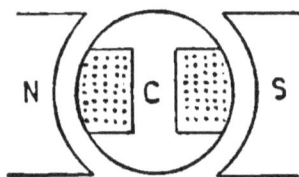

Fig. 24 B.

between the poles of a powerful magnet it becomes greatly heated by the currents which are induced in the metal.   Tyndall melted fusible metal in a copper tube rotated rapidly in a strong field. The induced currents are in this case in the metal and not in the coil and are generally known as " Foucault " currents.   The heat evolved depends on the form and material of the armature and on its velocity, and heat from this cause has always been a serious objection to the early form of Siemens' armature.   The next step was to have two armatures in the same machine ; one rotating between the poles of permanent magnets inducing a current which passed through the coils of a large electromagnet, between the poles of which the other armature was placed.   The adoption of electro-magnets greatly intensified the field, and as the current causing them was also generated by the machine, a great gain of power and efficiency was secured.   The machines of Ladd and Wilde were of this type.   The above were called magneto-electric machines, as they all depended on permanent magnets to start them, but in 1867 the permanent magnets were suppressed, the current from the armature passing through the coils of the electro-magnet, the " field " coils being in series with the external circuit and armature.   When the machine was stopped it was found that the cores of the electro-magnets possessed sufficient residual magnetism to induce a current in the armature when it was started again, and this current once induced, strengthened the electro-magnets and in turn induced more current. Machines of this type were called *dynamo-electric* or simply *dynamo* machines. The advantages they possess over the magneto-electric are greater power, the field being stronger; and greater economy, the magnets being of wrought iron instead of steel. The distinction between magneto and dynamo machines is now so slight, by the introduction of a variety of new types, that it is hardly worth preserving. As has already been shown, the energy of the induced current is derivable from the energy expended in moving the coil, and Prof. Thompson in his Cantor lectures has given a broad definition of a dynamo machine as "a machine for converting energy in the form of dynamical power into energy in the form of electric currents by the operation of setting conductors (usually in the form of coils of copper wire) to rotate in a magnetic field." Accepting this definition, the theory of the dynamo is best understood by a reference to the laws of electro-magnetic induction already examined. The induced electromotive force in a conductor moving in a magnetic field is $E = -\dfrac{dN}{dt}$. As an illustration, examine the case of a coil spinning in a uniform field, and the application of the formula to dynamos may be considered later. Suppose a coil, as in Figure 25, rotating on a vertical axis, the lines of force passing from the reader down through the paper perpendicularly. It incloses a maximum number of lines of force, and if rotated so that the right-hand edge comes to the front, while the left-hand goes behind the paper it will inclose a constantly decreasing number of lines, and a positive current will be induced. The E. M. F. will at first be small, as the rate of

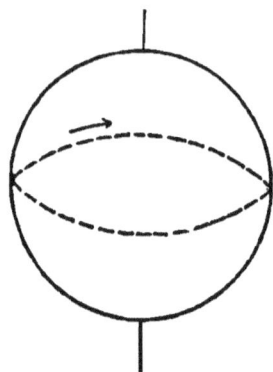

change is small, the edges of the coil moving almost along the lines of force. The rate will gradually increase until the coil has moved through one quadrant and is edge on to the observer, when, as the motion of the edges is at right-angles to the lines, the rate, and consequently the E. M. F., is a maximum. In this position the coil incloses no lines of force, and during the second quadrant it will move, inclosing an increasing number, and inducing, therefore, an inverse current. But the side of the coil now seen is the opposite to that in view during the first quadrant, and the inverse current is, therefore, in the same absolute direction in the coil as the former direct current. During the second quadrant the rate and E. M. F. decrease, becoming a minimum when the coil has completed a half revolution and is again in the plane of the paper. On entering the third quadrant, the number of lines inclosed decreases, and a direct current is induced ; but as the same side of the coil is presented to the observer as in the second, the direction of the current is reversed in the coil. In the fourth quadrant the number of inclosed lines increases, but the other side of the coil is toward the observer, so that the absolute direction of the current is the same as in the third. The general direction of the current is, therefore, downward in that part of the coil in front of the paper, and upward in the other half ; but as regards the coil itself, the direction of the current changes twice in every revolution, the point of change being where the circuit incloses the maximum number of lines of force. By the use of a commutator which shifts its connections at this point of the revolution, the current

Fig. 25.

may be made to flow in one direction in the exterior circuit.

Considering this coil as the armature of a dynamo machine, it is apparent that the current could be kept in one direction in the exterior circuit, but would be of varying strength. If another coil were fixed on the same axis but at right angles to the first, its E. M. F. would be a maximum when that of the first was a minimum, making the current in the external circuit more nearly uniform. By increasing the number of coils a practically uniform current could be obtained, but at the expense of a very complicated commutator.

68. Electromotive Force.

From the formula $E = -\dfrac{dN}{dt}$ it is evident, 1. That the E. M. F. varies as the *rate of change of the field.* For a constant time dt, the rate varies as the number of lines taken out or introduced, and the field should therefore be intense. Mere intensity is not, however, enough, as a coil could be moved in the most intense *uniform* field without inducing any current. The field must be so arranged that the coil either passes from a maximum positive to maximum negative, or what amounts to the same thing, that it rotates in a constant field, the lines being alternately added and subtracted. In this case the number of lines should be a maximum or the intensity cf the field as great as possible.

Perfect working in a dynamo requires a constant change of E. M. F., and consequently a constant rate. If a large coil revolves between the poles of two bar magnets, and no iron is present to modify the distribution of the lines of force in the field, the greater part pass direct

7

between the poles, and are cut during a small part of the revolution of the coil, during which time the rate and induced E. M. F. are high, but in other parts of the revolution the rate is very small. The available E. M. F. is induced suddenly, but the sudden creation of a current causes high self-induction and temporary strong extra currents, which in a dynamo are not only prejudicial but dangerous, on account of the high E. M. F. they may have. Idle wire in the armature also reduces the current. It is therefore desirable to prevent a concentration of the lines of force in a small part of the field. In a coil rotating in a uniform field, the advantage of the constant rate is attained by the change of the number of lines inclosed in the ratio of the sine of the angle between the plane of the coil and the direction of the lines, and as a field tends to become uniform would this advantage be gained.

To secure uniformity and prevent concentration large pieces of iron called pole pieces are frequently attached to the poles, partially encircling the armature. By using long magnets and heavy pole pieces, the field may be made nearly uniform. An important modification of the field arises from the lines of force due to the current in the armature. Thus in

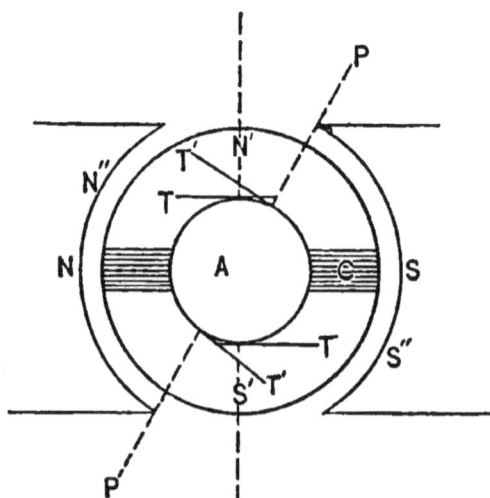

Fig. 26.

Fig. 26, representing a cross section of a Siemens armature, *A* being the end of the commutator and *TT* the

commutator brushes ; the lines of force of the field ordinarily pass from N to S in approximately straight lines.
When the armature is in revolution, each coil in succession
has its maximum current when it is in the position C in
the figure, and the effect of the armature current is therefore to induce two poles in the rim of the armature at N'
and S'. The poles N and N' may be supposed to form a
resultant pole at N'' and S and S' at S'', and the general direction of the lines of force of the field is therefore
N'' S''. As shown in the discussion of the revolving coil,
the commutator brushes should make contact at the
neutral points, at right angles to the lines of force. If the
lines of force of the machine in motion were in the same
position as when at rest the brushes might remain at
TT in a line perpendicular to the lines of force NS. If
left in this position, however, it will be observed that
there is a constant succession of sparks at the brushes,
which evidently do not press at the neutral points. This
sparking may generally be suppressed by rotating the
brushes into positions $T'T'$. The explanation is simple :
they have been brought into a line PP perpendicular to
the changed direction $N''S''$ of the lines of force of the
field. The stronger the current, the stronger the induced
pole of the armature N', and the nearer N' is the resultant
pole N''. The stronger the current, therefore, the more
the brushes must be advanced.

The different types of dynamos vary principally in the
way in which the field is formed. The principal methods
are :

(1.) The magneto in which the field is due to permanent
magnets. (2.) The separately excited dynamo, a separate
machine being used, the current of which passes through
the field magnet coils of the generator. This possesses
the great advantage of having a constant field, and when

several machines are used in one place, one may be
used to actuate the field magnets of all the others. (3.)
The series dynamo, the field coils being in the main circuit. As the whole current of the machine passes around
the magnets, an intense field is produced, but with the
great disadvantage that any increase of resistance in the
external circuit weakens the field, and consequently the
E. M. F., just when a high E. M. F. is necessary to overcome the increased resistance. (4.) The shunt dynamo
has the field magnet coils in a shunt of the main circuit.
In this type an increased external resistance sends a
greater current through the magnet coils, causing a more
intense field and a higher E. M. F. By having the resistance in the magnet coils adjustable, a shunt dynamo may
be made to give a practically constant E. M. F., whatever
the external resistance may be. (5.) A mixture of the last
two types has been used for special purposes, and is
known as a series and shunt, or compound dynamo. It is
a shunt dynamo having in addition a number of coils of
wire on its field magnets in the main circuit. In a shunt
dynamo, if the external resistance is suddenly increased,
a greater part of the current flows around the field coils,
inducing a higher E. M. F. To keep this constant the
speed would have to be decreased. In a series dynamo,
however, an increase of external resistance diminishes the
E. M. F., and to keep it constant the speed must be increased. By making the magnet coils partly in series and
partly in a shunt circuit, it is, therefore, possible to keep
the E. M. F. practically constant at a certain speed within
wide changes of the external resistance. (6.) Some
machines have two coils on the armature, one of which
sends a current through the field coils, while the other is
in the external circuit. (7.) Alternate current machines
are machines giving currents first in one direction and

then in the opposite. Their field is caused generally by another machine, but some types send a part of their own current, rectified by a commutator, through the field coils.

2. The E. M. F. varies as the *velocity*.

This relation is almost absolutely true in magneto machines and in others having a constant field, but in the series dynamo, where the field is itself a function of the current, the rate of increase of E. M. F. is much greater than that of increase of velocity up to the point of saturation of the magnets, beyond which, the field being constant, the above relation holds. Its correctness is assumed in practice.

A great waste sometimes occurs from the commutator brushes not being adjustable. As already shown, the lines of force of the field are distorted by the current, and this distortion is greater as the velocity is increased. The brushes must, therefore, be advanced, or they will take off the current at the wrong time, involving a waste of energy and causing "sparking," injuring both commutator and brushes.

3. The E. M. F. varies as the *number of turns of wire* in the armature.

The formula $E = -\dfrac{dN}{dt}$ is derived from a conception

of the work done in moving a single coil in a magnetic field. If *n* coils were moved either one by one or all together, *n* times the work would be done, and *n* times the E. M. F. induced. Increasing the number of turns increases the internal resistance in the same ratio, but if the external resistance is large there is a gain by taking more turns of wire on the armature.

4. The E. M. F. is greatest when the coil cuts the lines

of force at right angles. The rate of change is then greatest, and hence the electromotive force.

5. The E. M. F. varies as the *area* of the coil. If the field were uniform the gain would be directly as the area. In any case it varies directly as the number of lines of force inclosed, and there is, therefore, generally speaking, an advantage in having the coil of large area.

In the five considerations on which the E. M. F. of induction depends, the only variable, after the machine is made, is the velocity. Several of the types referred to admit of a partial adjustment to meet changed circumstances, but in general a machine should be adapted to the work expected of it, and should not be expected to be efficient under very different conditions. Although the velocity may be easily varied, it cannot be indefinitely increased without mechanical injury.

69. Efficiency.

A dynamo has properly two efficiencies. As it is a vehicle for the transformation of mechanical into electrical energy its *gross efficiency* is the ratio of the current energy to the mechanical energy actually applied to the machine. If an engine developing 16 H. P., is working a dynamo, two H. P. being lost in transmission to the dynamo, in friction and in overcoming the inertia of the engine, only 14 H. P. are actually applied to turn the armature. If in this case the electrical energy developed by the dynamo, was 10 H. P. the gross efficiency is $\frac{10}{14}$, or 71 per cent. A good dynamo possesses an efficiency of from 90 to 95 per cent. when working under the most favorable conditions, and therefore far surpasses any other machine in its capacity for transforming energy.

The ordinary use of the dynamo is to produce light. Whatever its use may be, all the electrical energy not utilized in producing the desired result is practically wasted. The *net efficiency* is the ratio of the electrical energy in the external circuit to the mechanical energy applied to the armature. If in the above case only 6 H. P. existed in the external circuit, the *net* efficiency $= \frac{6}{14}$, or 44 per cent. The distribution of the energy in the circuit is one of the most important problems relating to dynamos. The total work in circuit is, by Note 55, C^2Rt, R being the total resistance, consisting of internal r, and external l. The work is then $C^2r + C^2l$, and the ratio of the work done in the machine to that in the external circuit is $\dfrac{C^2r}{C^2l} = \dfrac{r}{l}$. The fraction of the total electrical energy in the external circuit is similarly $\dfrac{C^2l}{C^2R}$

$= \dfrac{l}{R}$. That this proportion should be great, l must be nearly equal to R, or in other words, the resistance of the machine must be small compared with that of the circuit. From the above the important relation is evident, *that the distribution of energy in an electrical circuit is determined by the distribution of the resistances in circuit.*

The efficiencies may now be calculated. Let a = resistance of armature, f that of the field coils, l of the external circuit, and R be the total resistance. E is the electromotive force, and C the current in circuit.

The *gross* efficiency of a series dynamo

$$= \frac{\textit{Work of current}}{\textit{H. P. applied}}.$$

This by Note 55 is

$$\frac{\dfrac{C^2R}{746}}{H.\,P.\,applied} = \frac{C^2R}{746 \times H.\,P.\,applied}.$$

This energy is given off in all parts of the circuit, that in the armature being C^2a (in watts), that in the field coils C^2f, and in the external circuit C^2l.

The *net* efficiency is

$$\frac{\textit{Work in external circuit}}{H.\,P.\,applied};$$

and this is

$$\frac{\dfrac{C^2l}{746}}{H.\,P.\,applied} = \frac{C^2l}{746 \times H.\,P.\,applied}.$$

The energy wasted as heat in the machine is $C^2(a+f)$ and the ratio of energy wasted is $\dfrac{a+f}{a+f+l}$. That this may be small the internal resistance $(a+f)$ must be small in comparison with the external. The energy wasted takes the form of heat, and is thus not only wasted but directly harmful, as heating of the machine increases its resistance and thereby increases the ratio of wasted energy.

In the shunt dynamo the relations are more complex, as the current in the various branches of the circuit is different.

Let a = armature resistance and A armature current,

f = field coil   "   " F current in field coil,

l = external   "   " L "   " external circuit,

$$R = \text{Total resistance} = a + \frac{lf}{l+f}.$$

Work in armature $= A^2a$, in field coils F^2f, and in external circuit L^2l.

Total electrical energy $= A^2R = A^2\left(a + \dfrac{lf}{l+f}\right)$.

Gross efficiency $=$

$$\frac{\dfrac{A^2\left(a + \dfrac{lf}{l+f}\right)}{746}}{H.\,P.\,applied} = \frac{A^2\left(a + \dfrac{lf}{l+f}\right)}{746 \times H.\,P.\,applied}.$$

To find the net efficiency, the ratio of the electrical energy utilized is

$$\frac{L^2l}{A^2\left(a + \dfrac{lf}{l+f}\right)} = \frac{\left(A\cdot\dfrac{f}{f+l}\right)^2 l}{A^2\left(a + \dfrac{lf}{l+f}\right)} = \frac{f^2l}{a(f+l)^2+f^2l+fl^2};$$

and by multiplying this into the value of the gross efficiency, previously obtained, the product is the net efficiency.

The above expression contains only resistances. If L is measured the net efficiency is evidently

$$\frac{L^2l}{746 \times H.\,P.\,\text{applied}}$$

70. Electromotive Force in Circuit.

In using CE to calculate the electrical energy in any case, E must not be taken as the difference of potential at the machine terminals. Calling this difference of potential E', and considering the case of the series dynamo as being more simple, we have from Kirchhoff's second law, $E' = Cl$, l being the external resistance,

or $\quad C = \dfrac{E'}{l}$, but from Ohm's Law $C = \dfrac{E}{l + r}$

$\therefore \dfrac{E'}{l} = \dfrac{E}{l + r}$; or $E = \dfrac{E'(l + r)}{l}$.

Whenever E' is measured E must be calculated if the total electrical energy is to be computed. The product CE' is evidently the energy in the external circuit, and is less than the total energy by the quantity C^2r expended in the machine. The total energy is, therefore, $CE' + C^2r = C^2(l + r)$. If the resistances are all known the total energy may be calculated from the last formula without any risk of error.

It has been shown by Sir William Thomson that in shunt dynamos the best results are obtained when the external resistance is a mean proportional between the resistance of the magnet coils and that of the armature, the latter being small in comparison with the resistance of the magnet coils.

71. Siemens' Machine (§ 409, Fig. 151).

This is a shunt dynamo. The armature is similar in shape to Siemens' armature already described, being a cylindrical drum, but having several coils coiled on it lengthwise instead of one. There are as many divisions of the commutator as there are coils, the divisions being longitudinal. An eight-coil machine has, therefore, its commutator ring divided into eight segments, to each of which connect the ends of two coils. The other ends of these coils are connected to other commutator divisions, so that the eight coils are all in a continuous circuit between the commutator brushes, so wound that in all eight the current at any given instant flows in the same direction. In some coils the E. M. F. is greater than in others, but as there are so many, the total E. M. F. of all in series varies but slightly from time to time, and the current is, therefore, practically constant. By placing the commutator brushes opposite each other, they are in contact with points of the circuit differing most widely in potential, and a permanent difference of potential is therefore maintained between the terminals of the machine. The induction of the current in any one coil is analogous to that in the coil described in Note 67.

72. The Gramme Machine (§ 410, Fig. 153).

The Gramme is generally a series dynamo, although sometimes separately excited, and sometimes having its field coils excited by a separate armature coil. The armature is a ring of soft iron wire, widened till it might be considered a short hollow cylinder. Around this ring are coiled a great number of armature coils, as shown in Fig. 152, the ends of the coils being brought to divisions of the commutator. The commutator consists of a number of plates radially arranged around the axis of the armature, and insulated from each other. The commutator divisions are seen on the right of the armature in Fig. 153, and correspond in number to the armature coils, which are connected through them in one continuous circuit.

The action of the Gramme may be easily understood from the rules of Note 59. In Fig. 152 the positive direction of the lines of force is from N to S, the lines entering the ring opposite N, and dividing, running through each half of the ring to that part opposite S, where they leave the ring and pass to S. The poles N and S cannot be considered as points, and the lines, therefore, enter the ring all along its lower portion (as shown in the figure) and emerge along the upper part. A coil in the position E'' has, therefore, the maximum number passing through its plane. If, now, the armature is rotated, so that E'' passes towards E, it continually incloses a decreasing number of lines of force, and a direct current viewed from N is induced. The E. M. F. varying as the rate of change, is zero at E'' and a maximum at a point opposite S, where the coil cuts all the lines at right angles. As the rotation of the armature continues, the coil after leaving E incloses an increasing number of lines of force, and the current is therefore inverse as viewed from N. But from E to E' the side of the coil viewed is the opposite of that seen from E'' to E, and the inverse current in the quadrant from E to E' is therefore in the same absolute direction in the coil as the direct from E'' to E. Throughout the half revolution from E'' to E', therefore, the induced current flows in the same direction, being strongest when the coil is nearest the pole S. By connecting all the coils in series, the E. M. F. in the circuit becomes the sum of all in the individual coils, and as these occupy all possible positions at any instant, the total electromotive force is constant, the machine thus yielding an almost absolutely constant current.

The action during the other half of the revolution may be traced in the same way. The coil in moving from E' to N incloses a decreasing number, inducing a direct current, which is opposite in direction to that in the quadrant from E to E'. If, therefore, in the latter the current had flowed away from the point E' towards E, it would in $E'N$ flow away from E' towards N, and although the absolute direction of the currents is different they combine to lower the potential of E'. During the quadrant between N and E'', the coil incloses an increasing number, and consequently has an inverse current induced, but this inverse current is in the same absolute direction as the direct in the preceding quadrant. If, therefore, throughout the upper half of the revolution the current flows away from E', it will in the lower half of the revolution flow away from E' also. Throughout the whole revolution the effect is to raise the potential of E'' and lower that of E', and if brushes touch the commutator at these points they will possess a difference of potential which may be utilized in the production of a current through an external circuit. The Gramme machine has been the subject of much investigation, and its action has been variously explained. The most general explanation in any case of electromagnetic induction is that obtained from a consideration of the lines of force, and this is the one adopted by Prof. Thompson, which has only been given here in slightly greater detail.

The armature cylinder is made of soft iron wire, both to facilitate the rapid magnetization and demagnetization, and to prevent heating from the Foucault currents which would take place if solid metal were used. The change of direction of the lines of force of the field by those due to the current is frequently very marked in the Gramme machine, M. Breguet having found it necessary to advance the commutator brushes 70° when working with a Gramme at 1770 revolutions. As the internal resistance of the Gramme is generally small, it is specially adapted for working a single powerful arc light, while the steadiness of its current renders it well adapted for incandescent lighting.

73. The Brush Machine (§ 411).

This machine has received its main development in the United States, but is now extensively used throughout the world. It contains many peculiar features, and is distinctly a separate type, although frequently alluded to, especially by French authorities, as a modification of the Gramme. The general appearance of the machine is shown in Fig. 27.

The first noticeable peculiarity is in the disposition of

Fig. 27.

the four field magnets, which are placed so that the armature coils pass between similar poles. The magnets are
oval in cross section, and are furnished with large pole
pieces, approaching very closely on each side to the armature. The armature is a soft iron disc, with deep circular
furrows cut in its sides to break the continuity of the surface and thus prevent the heating of the metal by the induction of Foucault currents. On the periphery of the
armature of the small machine there are eight coils, the
two coils diametrically opposite being in one, but coiled in
opposite directions (See Fig. 30), so as to act in unison in
the induction of currents. The coils project from the armature as seen in Fig. 27, the reason assigned being, that
the fanning of the air thus caused prevents overheating.

The commutator consists of four rings each split into
four segments. A cross sec-
tion of one of the rings is as in
Fig. 28, the two ends of one
pair of coils being connected
to the segments marked I, I,
which are insulated from the
segments 2, 2. When the
brushes of the commutator
touch the latter the coils are
cut out of circuit. These cutting out segments in the dif-

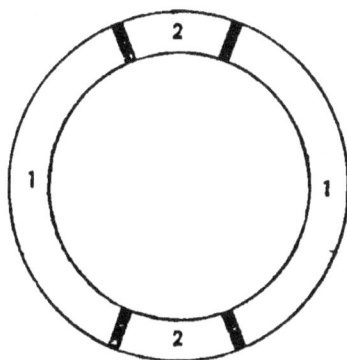

Fig. 28.

ferent rings of the commutator are so placed that at every
instant one coil is cut out ; the connections being made
so that a coil is not in circuit in that part of the revolution when no current is being induced in it. Each of
the four brushes presses on the commutator rings of two
coils not adjacent. Numbering the coils on the armature
I, 2, 3 and 4 (Fig. 30) in order, the brushes B^1 and B^4 are in
circuit with coils I and 3 and B^2 and B^3 with coils 2 and 4.

As the armature revolves each coil successively passes through all parts of the field. When a coil is midway between the dissimilar magnet poles, at the highest point of its revolution, the number of lines of force inclosed is a maximum, but changes so slowly that for this portion of the revolution the induced current is but small, and the coil is, therefore, cut out. As the coil approaches the large pole pieces and passes between them the rate changes rapidly. If a piece of soft iron be placed between two powerful similar magnet poles, the lines of force pass into it almost parallel on each side, and a coil moved along the bar cutting them perpendicularly, has a high rate of change in the number of lines inclosed, and consequently a high electromotive force induced. Thus in Figure 29, as nearly all the lines of both poles pass through the soft iron between a and b, the coil A in moving within that region experiences but little change in the number inclosed, but as it approaches either end the rate of change is very great. The electromotive force is thus in-

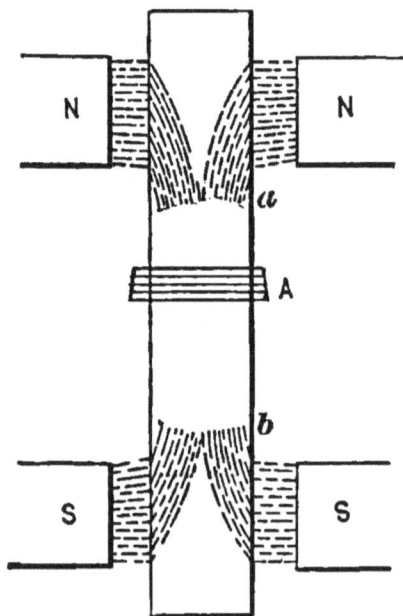

Fig. 29.

duced somewhat suddenly, but the efficacy of the peculiar arrangement of poles for the induction of a high electromotive force is evident. A comparison between Figs. 29 and 30 shows this to be nearly the condition existing in the Brush machine. The latter figure is a plan

of the machine.   As each pair of coils is connected to a
separate commutator ring, the study of the connections is
necessary to understand the complete working.   In the
figure, L represents a lamp in the external circuit. B^1, B^2,
B^3 and B^4 are the commutator brushes, and the rings are

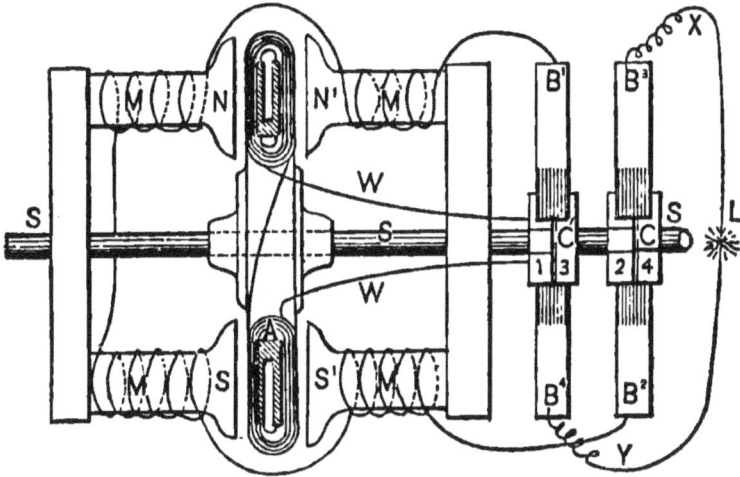

Fig. 30.

numbered 1, 2, 3, 4, as illustrating the way in which the
ends of the coils, taken in regular order around the armature, are connected at the commutator.   The currents induced in the several coils at any one instant will have the
following circuits, coil 4 being supposed to be cut out :

Coil 1—1, B^4, L, X, B^3, 2, B^2, M, B^1.

Coil 2—   2, B^2, M, B^1 $<^1_3>$ B^4, $L X$, B^3.

Coil 3—3, B^4, L, X, B^3, 2, B^2, M, B^1.

These paths are the same except in the armature coils
and the resultant current will have the path

$$B^1 <^1_3> B^4, L, X, B^3, 2 B^2, M, B^1.$$

An instant later coil 4 will be in circuit and 1 cut out.
The resultant current then flows

$$B^1, 3, B^4, L, X, B^3 <^2_4> B^2, M, B^1, 3.$$

8

The E. M. F. in circuit is evidently that due to two coils in series, and the internal resistance of the machine is diminished by the fact that there are always two of the four armature coils in parallel arc.

The gross efficiency of the Brush machine is lower than that of some others, but it possesses the great advantage of yielding so high an electromotive force as to be able to burn forty arc lights in series, a feat which no other machine can accomplish. As there are only two coils in series at one time, the resultant electromotive force is far from constant, and the fluctuations are so great as to utterly unfit the machine for incandescent lighting or other purposes requiring a constant current. The E. M. F. of the largest Brush machine is 2000 volts and the current about 10 ampères.

74. Edison Machine.

The Edison machine (¿ 411) is a shunt dynamo. Its chief peculiarities are its long cylindrical magnets ending in remarkably heavy pole pieces almost encircling the armature, and the peculiar construction of the armature itself. Theoretical investigation and experiment both point to long cylindrical magnets as most efficient ; and in a shunt dynamo, in which there is a perpetual endeavor for a permanent adjustment of the strength of the field to the necessities of the case, it is advantageous to have magnets of considerable mass, as the change of field brought about by a variation in the strength of the field current is thus made more gradual. The large pole pieces tend to make the field more uniform, and thus act to secure a constant rate or a uniform change of electromotive force throughout the rotation.

Edison calls his large machine a "steam dynamo," the engine and dynamo being on the same bed-plate. It is specially designed for use at a central station to supply power or work incandescent lights throughout a district of a city. As established in New York, the whole weight of dynamo and engine is nearly thirty tons, sixteen of which are in the magnets and pole pieces. The core of the armature is made up of sheet iron discs, separated from each other by tissue paper and bolted together. This prevents heat currents. Instead of wire, the armature circuit is made of heavy copper bars, each bar being insulated from the next and from the iron core by an air space. The bars are connected together at each end of the armature by copper discs, there being half as many discs at each end as there are bars. Each disc has lugs formed on it on opposite edges, to which two bars are connected, and the whole being bolted together, the bars and discs form one continuous circuit of wonderfully low resistance, the total armature resistance of a machine sent to London being .0032 of an ohm. This very low resistance is necessary from the fact that the machine is intended to work 1300 incandescent lights, each of about 137 ohms, in parallel arc. The external resistance would, therefore, be only .095 ohms. As the number of lamps in circuit changes, the resistance in the magnet coils, which are in a shunt of the main circuit, is regulated so as to keep a practically constant electromotive force, and each lamp then burns with the same intensity under all conditions. Edison's large machine gives an E. M. F. of 110 volts and an ordinary current of 1000 ampères.

75. Alternate Current Machines.

Alternate current machines have been used in Europe to a considerable extent for incandescent lamps and the Jablochkoff and other candles. Almost any machine yields alternate currents if used without a commutator, but most alternate current machines have a large number of armature coils which pass between the poles of a system of opposed magnets, so arranged that the positive lines of force in the field pass alternately through the coils in opposite directions, many machines inducing fifteen or twenty currents in each direction in every revolution. By placing a number of coils in series, so disposed that the induced current in all is in the same direction at any instant, a high electromotive force may be induced. Many machines have the connections of the coils adjustable, so that they may be arranged either in series or in arc, thus permitting an adaptation to the requirements of the external circuit. An objection to alternate current machines is that the frequent reversals of current induce extra currents of so high electromotive force as to be dangerous. The more coils there are in the machine, the higher the coefficient of self-induction, and the greater the velocity the greater the induced electromotive force, so that the extra current may be much stronger than the normal current of the machine. This disadvantage exists with many continuous current dynamos, particularly the Brush, the electromotive force of which is, as already stated, not only high but also very variable. Every change of the normal current induces an extra current, of greater strength as the revolution of the machine is more rapid. Alternate current machines are less economical than those generating a continuous current.

VIII. ELECTRIC MOTORS.

76. General Principles (§ 375).

In describing the action of Ritchie's electric motor, Prof. Thompson abandons the method pursued elsewhere in his book, that of a consideration of the lines of force in the field, and adopts another less general explanation, that of the mutual action of magnet poles. If the core of the coils CD were of some non-magnetic substance the description given would not apply, although the motor would still work. The two coils C and D in Fig. 141 are practically one, and this will in any magnetic field tend to place itself so as to bring its own lines of force in the same direction as those of the field. The lines of the field pass from N to S, and if those due to the current in the coil are opposite in direction, the coil will tend to rotate into a position of equilibrium, but just before this is attained the rotation shifts the connections of the coil in the mercury cups, the current changes, its lines are again opposite to those of the field and the coil continues its rotation through another semicircle. Owing to the continued shifting of the direction of the current, the coil is perpetually in unstable equilibrium, and the rotation is continuous in the endeavor to attain equilibrium. As shown in Note 36, the work done by a coil in a half revolution is $2CHA$, and if the coil is of n turns this becomes $2CnHA$. This experiment well illustrates many of the principles of electric motors, and particularly that of the commutator.

77. Electric Transmission of Power to a Distance (§ 376).

As illustrated more fully in Note 78, mechanical energy may, by the use of a dynamo machine, be converted into electrical energy, and be transferred into mechanical energy again by another dynamo at a distance. The value of this fact depends on circumstances. There are very many cases in which power thus obtained, due primarily to a water-fall at a distance, would, in spite of the great loss in transmission, be more economical than the same power generated from steam on the spot. If incandescent lighting becomes an established system in cities, as started by Edison in New York, the same wires which work the lamps at night may transmit power by day, and small motors may be used on the lamp circuits. The readiness with which electric energy in the form of a current may be subdivided, offers great advantages for its distribution in small amounts over the system of distribution of steam power by a complicated system of shafting and belting. Another point in which electric motors may be used is in electric railroads. A stationary steam engine is vastly more economical than a locomotive, both in wear and tear and in consumption of fuel and water. Generating the required power by stationary engines, this economy may be pushed to the utmost by the adoption of large engines of the most approved type; and as an electric motor may be made of great power but of little weight, the injury to rolling stock and road bed resulting from the use of heavy locomotives would be prevented. Stationary engines working dynamos may be placed as needed and the current generated be transmitted through the rails to the motor. Several such roads have been constructed for short distances, and the question of their development is merely one of commercial economy.

78. Theory of Electric Motors (§ 377).

The theory of electric motors first propounded by Jacobi has of late received mathematical development from others, and has been the subject of much experiment. The underlying principle is that referred to in Note 76, that if a current be passed through a coil, free to move in a magnetic field, the coil will move into a certain position, and in moving is capable of doing work. This principle is the converse of that underlying the action of the dynamo machine, and it is therefore easy to see that a machine which will generate currents when worked, will, on the other hand, work when a current is sent through it, the rotation as a motor being in the opposite direction to that as a generator of current. By using two machines in the same circuit, the current generated by one will cause the other to work. The commercial value of the fact is determined by the cost of the power given off by the second machine.

If C be the current in circuit and E the E. M. F. of the generator—not the difference of potential at the terminals of the generator (See Note 70)—the electrical work of the generator is CE. The motor in rotating backwards generates a current in the circuit in the opposite direction to that of the generator, thus *diminishing* the current in circuit. This follows directly from the principle of conservation of energy. If ε is the back electromotive force of the motor, the electrical work of the motor is $C\varepsilon$. The mechanical work given off by the motor can never equal this, as the motor is not a perfect vehicle for the transmission of electrical into mechanical energy, but in calculation the motor may be assumed to be perfect, and corrections applied to the final results.

The ratio of the work of the motor to that of the generator is

$$\frac{C\varepsilon}{CE} = \frac{\varepsilon}{E} \quad \ldots \ldots \quad (1)$$

or the return is the ratio of the electromotive forces of the two machines. If these are exactly similar and working with fields of equal intensity, a condition not holding in practice, the ratio of ε to E is that of the velocities of the two machines.

As the motor generates an inverse current, the current in circuit is

$$C = \frac{E - \varepsilon}{R} \quad \ldots \ldots \quad (2)$$

R being the total resistance.

The energy of the generator is expended partly as heat in the circuit, and partly as work in the motor

or
$$CE = C^2R + \text{work} \quad \ldots \ldots \quad (3)$$
$$\text{Work} = CE - C^2R.$$

Differentiating for a maximum,

$$\cdot \frac{dw}{dC} = E - 2CR = 0 ;$$

$$\text{or } C = \frac{E}{2R} \quad \ldots \ldots \quad (4)$$

The maximum work is, therefore, done by the motor when its velocity of rotation is such as to reduce the current in circuit to one-half that due under Ohm's Law to the electromotive force of the generator and the resistance in circuit. This is the case when $\varepsilon = \frac{1}{2} E$. The return is then by (1), $\frac{\varepsilon}{E} = \frac{1}{2}$.

From these equations the conditions of economy are deducible.   If the only consideration is that the motor should do the most work, equation (4) gives the condition that its back electromotive force should be one-half that of the generator.   If, however, the governing condition is that the motor should work as economically as possible, (1) indicates that the electromotive force of the motor should nearly equal that of the generator.   The return is, then, greater than one-half, but (2) shows that the current is reduced, that CE, the work of the generator, is also diminished and that consequently a greater proportion of a smaller amount of work is transmitted.   The governing consideration is whether the motor should do as much work as possible, regardless of cost, or work with the greatest economy, regardless of the amount of work done.

These conditions are of the greatest importance, but are somewhat difficult to reconcile.   Prof. Thompson has devised a graphic illustration which presents them very clearly.   Draw AB to represent E, the electromotive force of the generator, and on it construct a square, $ADCB$.   On AB lay off from B, BF to represent proportionally ε, the electromotive force of the motor, and on BF construct a square $BLGF$.   Through G draw FH parallel to BC and KL parallel to AB.

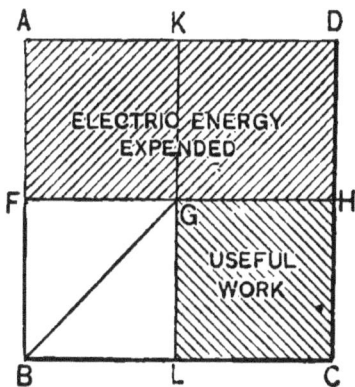

Fig. 31.

Then the work of the generator is CE $\dfrac{E(E-\varepsilon)}{R}$, the

work of the motor is $C\varepsilon = \dfrac{\varepsilon(E-\varepsilon)}{R}$.

The return $= \dfrac{C\varepsilon}{CE} = \dfrac{\varepsilon\,(E - \varepsilon)}{E\,(E - \varepsilon)}$.

But $\varepsilon\,(E - \varepsilon)$ is the area of $GLCH$, and $E\,(E - \varepsilon)$ is the area of $AFHD$. These areas are, therefore, marked as "electric energy expended" and "useful work," and the ratio of the two areas is the return. From the construction the point G will always fall on the diagonal BD, approaching D more nearly as F approaches A. The area $GLCH$ corresponding to the useful work will, therefore, always be inscribed within the triangle BCD, and will have its maximum value when it is a square as in Fig. 31. G is then midway between B and D, and the area of $GLCH$ is one-half that of $AFHD$. From similar triangles, ε is also one-half of E. This demonstrates the case of maximum work; that of maximum efficiency is evident from Fig. 32. The lettering and construction are the same, but the value of ε has been increased. The ratio of $GLCH$ to $AFHD$ is greater, although each area is less than in the preceding case. A greater ratio of a less amount of energy is thus shown to be transmitted. In the last

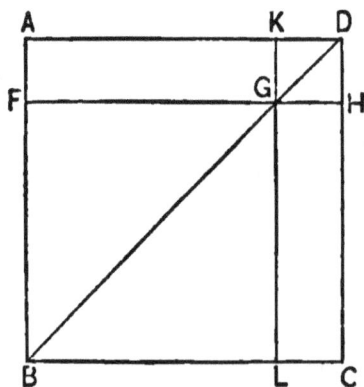

Fig. 32.

figure the area $GLCH$ is by geometry equal to $AKGF$, and the square $KDHG$, the difference between them, represents, therefore, the factor C^2R, or the loss by heat. The smaller this loss the greater the return. Any desired return may be calculated by making $GHCL$ the required fraction of $ADHF$. If this is to be 90 per cent. $KDHG$ must be $\frac{1}{10}$ of $ADHF$, and this may be secured by making

$DH \frac{1}{10}$ of DC, or what is the same thing $BF \frac{9}{10}$ of AB.
The geometrical construction, therefore, gives the same
result as that already obtained, that the return is the ratio
of the electromotive forces, or that $\dfrac{\varepsilon}{E}$ should be $\frac{9}{10}$.

The work done by the generator being CE the question
arises whether it is best to increase the energy by using
a stronger current or a higher electromotive force. The
loss by heat is C^2R, and this by (2) is equal to

$$\frac{(E - \varepsilon)^2}{R} .$$

If now E and ε are both increased by the addition of the
same numerical quantity, the difference $(E - \varepsilon)$, and consequently the loss by heat, is the same. Calling the new
values E' and ε' the amount of work done is easily calculated. The work done by the generator is now

$$\frac{E'(E' - \varepsilon')}{R} = \frac{E'(E - \varepsilon)}{R} \quad \ldots \ldots \quad (5)$$

and that by the motor,

$$\frac{\varepsilon'(E' - \varepsilon')}{R} = \frac{\varepsilon'(E - \varepsilon)}{R} \quad \ldots \ldots \quad (6)$$

The original work of the generator was $\dfrac{E(E - \varepsilon)}{R}$. (7)

and of the motor $\dfrac{\varepsilon(E - \varepsilon)}{R}$ $\quad \ldots \ldots \ldots \ldots$ (8)

As E' and ε' are greater respectively than E and ε, a
comparison of (5) with (7) and (6) with (8) shows that
more work has been done by the generator and more
by the motor, while the loss by heat was the same.
There is, therefore, clearly an economy in using high

electromotive forces both in generator and motor.   The
use of high electromotive force is, however, more dangerous and necessitates better insulation.

The above sets forth the general conditions of transmission of energy by electricity.   A few deductions are
easily made.   Solving (3)

$$C = \frac{E \pm \sqrt{E^2 - 4RW}}{2R} \quad \cdot \quad \cdot \quad \cdot \quad \cdot \quad (9)$$

Equating (9) and (2),

$$\varepsilon = \frac{E \mp \sqrt{E^2 - 4RW}}{2} \quad \cdot \quad \cdot \quad \cdot \quad \cdot \quad (10)$$

From (10) and (1)

$$\text{Return} = \frac{\varepsilon}{E} = \frac{1 + \sqrt{1 - \frac{4RW}{E^2}}}{2} \quad \cdot \quad \cdot \quad \cdot \quad \cdot \quad \cdot \quad (11)$$

(11) shows that the return is not independent of the
amount of work done by the motor, the return diminishing, other things equal, as the work done increases.   If,
however, the return is wished to be the same, it may
be secured by making the work done vary inversely as
the resistance through which it is transmitted.

Denoting the return, or $\dfrac{\varepsilon}{E}$, by K, the following equa-

tions are readily deduced.

$$\text{Work of generator} = \frac{E(E - \varepsilon)}{R} = (1 - K) \cdot \frac{E^2}{R} \quad \cdot \quad (12)$$

$$\text{Work of motor} = KCE = K(1 - K) \cdot \frac{E^2}{R} \quad \cdot \quad \cdot \quad \cdot \quad (13)$$

$$\text{Loss in heat} = \frac{(E - \varepsilon)^2}{R} = (1 - K)^2 \cdot \frac{E^2}{R} \quad \cdots \quad (14)$$

These equations, due to Marcel Deprez, show that work and loss by heat remain constant whatever the resistance in circuit may be, if E is made to vary as $\sqrt{R}$, that $\frac{E^2}{R}$ may be constant. By an increase, therefore, of the electromotive force of the generator, the same amount of work may be transmitted to a greater distance.

79. Modifications of Theory in Practice.

The preceding demonstration is entirely theoretical, and although the general conditions hold, they are much modified in practice, the modifications, moreover, being all unfavorable for economical results. Among the causes of error are the following : no dynamo machine or motor is a perfect device for transmitting energy. The work CE done by the first dynamo is less than that received by it, and the motor is unable to transfer the whole quantity $C\varepsilon$ into mechanical energy. If F represents the gross efficiency of the generator and f that of the motor, the work done *on* the generator is $\frac{CE}{F}$, and that done ˜by the motor is $f . C\varepsilon$. The commercial return is, therefore,

$$\frac{\textit{Mechanical work done by motor}}{\textit{Power applied to generator}} = \frac{fC\varepsilon}{\dfrac{CE}{F}} = f . F . \frac{\varepsilon}{E}.$$

Assuming for both generator and motor the gross efficiency of 90 per cent., the maximum commercial return is .81 of the power applied, the work done being then very small, or .9 of the theoretical return in the preceding equations. Another loss is that caused by leakage. If the insulation is not perfect, leakage occurs all along the line, and the current at the motor is less than the current at the generator. This loss is greater as the electromotive force is raised. If the motor is a self-exciting dynamo, the weaker current causes a weaker field than that of the generator, so that if the two machines are exactly similar, the ratio of the electromotive forces are not the same as the ratio of the velocities. There may be other causes operating, but these are sufficient to show that in no case can the theoretical return be fully realized. Marcel Deprez has at different times made experiments in transmitting power to a distance. The most careful up to this date were made in Paris on March 4, 1883, when he succeeded in obtaining from a motor 4.439 H. P. (French), which had been transmitted through a resistance of 160 ohms, corresponding to about ten miles of ordinary telegraph wire. The power applied to the generator was 12.267 H. P., but the electrical work CE of the generator was only 9.751 H. P. Gramme machines of high resistance were used, the electromotive force of the generator being 2,480 volts and that of the motor 1,779. The electrical return was, therefore, 71.7 per cent., the commercial return 36.2 per cent.

That the electric transmission of energy may be a success, a generator is necessary which with a constant speed gives a constant electromotive force, whatever changes may take place in the external circuit. If a generator is to work one hundred motors, it must be able to work them all at once or a few at a time, without running risk of injury from sudden changes of the number in use. Machines, or combinations of machines, for this purpose have been invented, one type of which, compound dynamos (Note 68, 1), has already been referred to. It is also necessary that the motors should move at the same speed whether doing work or not.   Improvements in these two points will make the transmission of power a commercial success.

80. Peltier Effect (§ 380).

This effect is caused when a current flows from one metal to another, and is independent of the resistance. As stated on page 343, the heating varies directly as the current, and the junction which is heated by a current in one direction is cooled if the direction of the current is reversed. Letting P be the heat in joules produced at a junction of two metals per second by a current of one ampère :

$$\text{Total heat in joules} = C^2R \pm PC$$
$$= C\,(CR \pm P).$$

In the last equation the quantity in brackets and the term CR are both electromotive forces. P must, therefore, be also an electromotive force measured in volts, although commonly called the "coefficient of the Peltier effect."

By carefully measuring the change of temperature at a junction with the current alternately in opposite directions :

$$H_1 = C^2R + PC$$
$$H_2 = C^2R - PC$$
$$H_1 - H_2 = 2PC \qquad \text{or } P = \frac{H_1 - H_2}{2C}.$$

If the current is one ampère,

$$P = \frac{H_1 - H_2}{2}.$$

H_1 and H_2 are measured in joules, or in ergs × 10^7, hence the numerical value of P may readily be found in ergs.

The equation in the Elementary Lessons was written before the joule had come into use as a practical unit of heat, but is the same as the above, since the joule is equal to .24 of a water-gramme-degree centigrade thermal unit.

81. Secondary Batteries (§ 415).

It has been found that lead dioxide is highly electro-negative to metallic lead, the difference of potential between the two in dilute sulphuric acid being about 2.7 volts. These two substances are used in the secondary battery, partly on account of the high difference of potential they possess, but mainly on account of the facility with which the lead dioxide may be formed by electrolysis. The Planté cell consists of two plates of lead in dilute sulphuric acid. If a current is passed through the cell, the liquid is decomposed, hydrogen is evolved on the kathode and oxygen on the anode. The latter unites chemically with the lead, forming lead dioxide, and this being, as stated, highly electro negative to lead, if the original source of electricity be removed and the secondary cell short circuited a current will flow through the cell in the opposite direction to that of the charging current, and will in time deoxidize the negative plate. The cell is then discharged. The process of charging is, therefore, merely one of polarization, and the effect of the current which it is of the most importance to avoid in the ordinary cell is the basis of the utility of the secondary battery. Charged in this way, however, a Planté cell yields but little current. In practice the cell is charged as above, discharged and then charged in the opposite direction, and this alternate charging in opposite directions in time renders both plates spongy or cellular in texture, enabling the oxygen given up at the anode to more readily enter into combination with the lead and forming a dioxide layer of greater thickness.

The Faure and later secondary batteries are similar in principle to the Planté, but have the plates at the beginning coated with lead oxide. When, then, a current is passed the anode is oxidized to lead dioxide, and the kathode deoxidized to metallic lead by the hydrogen evolved. The cell does not need the preliminary treatment of the Planté, but is ready for use immediately after the first charging, but the chemical action is much more complex, and the cell is probably not so durable.

The term "storage of electricity" is frequently used in connection with secondary batteries, but is not strictly accurate, as the portion of the energy of the charging current which is stored in the cell is in the form of energy of chemical separation, and is again transformed into electrical energy when the circuit is closed. The utility of the secondary battery arises from the fact that the chemical action when the circuit is open, is not great, and that the cell may be used after an interval of a few days from the time it was charged. The objection to its use is that, in in the first place, only a portion of the electrical energy of the current can be stored in the cell as chemical energy, and secondly, that chemical action does take place to a certain extent when the cell is not in use, and that it cannot, therefore, store energy indefinitely. The chief deterioration arises from the formation of lead sulphate. A cell, moreover, wears out eventually and becomes practically useless. There are many ways in which secondary batteries may be of service, particularly in connection with dynamos in electric lighting, but the anticipations of their sphere of usefulness entertained shortly after the introduction of Faure's battery were exaggerated.

9

82. The Morse Alphabet (§ 425).

The alphabet printed on p. 397 is the international alphabet used in Europe and, in fact, everywhere except in the United States and Canada, where the code originally introduced by Morse is still in use. The international is probably the better, as it is more easy to distinguish combinations of letters and avoid mistakes, but it is extremely difficult to make any change in a code, however faulty it may be, when it has once come into use. The alphabet used in America is as follows :

A	– ⎯	T	⎯
B	⎯ – – –	U	– – ⎯
C	– ‧ – ‧ –	V	– – – ⎯
D	⎯ – –	W	– ⎯ ⎯
E	–	X	– ⎯ – –
F	– ⎯ –	Y	– – ‧ – –
G	⎯ ⎯ –	Z	– – – ‧ –
H	– – – –	&	– ‧ – – –
I	– –	1	– ⎯ ⎯ –
J	⎯ – ⎯ –	2	– – ⎯ – –
K	⎯ – ⎯	3	– – – ⎯ –
L	⎯⎯⎯	4	– – – – ⎯
M	⎯ ⎯	5	⎯ ⎯ ⎯
N	⎯ –	6	– – – – – –
O	– ‧ –	7	⎯ ⎯ – –
P	– – – – –	8	⎯ – – – –
Q	– – ⎯ –	9	⎯ – – ⎯
R	– ‧ – ‧	0	⎯⎯⎯
S	– – –		

83. American System of Telegraphy* (§ 426).

The European system is known as the "open circuit" system, the current flowing only when the key is depressed. Many inconstant cells like the Leclanché may, therefore, be used. In America, however, the current flows continuously when no message is passing. When an operator wishes to telegraph he first breaks the circuit by a switch attached to the key, and then makes the signals, the circuit being closed when the key is depressed. When not

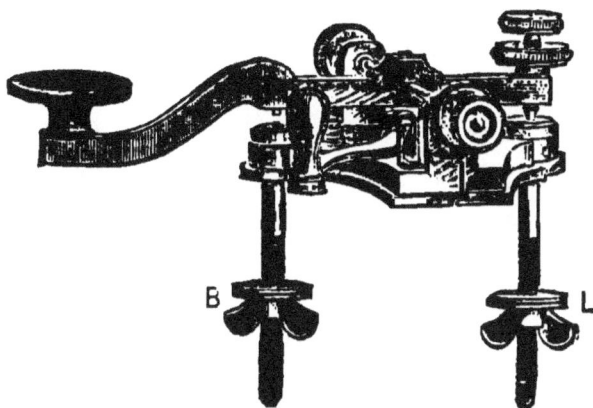

Fig. 33.—KEY.

telegraphing the switch must be closed or no signals can be made by any other operator on the line. The general appearance of the American key is shown in Fig. 33. The key is fastened to the table by the screws B and L, the former being insulated from the metal of the key, the latter in connection with it. One wire is connected to the metal of the key, generally at L, and the other clamped by B. The switch moves horizontally, and when pushed towards the left in the figure, makes contact with B

---

* Only a mere outline of the closed circuit system is here given. Full information may be found in books on telegraphy, the best being probably Prescott's " Electricity and the Electric Telegraph."

and connects it with L. When pushed to the right the circuit is open, and is closed only when the key is depressed and contact made with the head of the screw B.

The general arrangement of apparatus at a way station is shown in Fig. 34. The current entering by the line wire on the right first passes through the key K, the switch

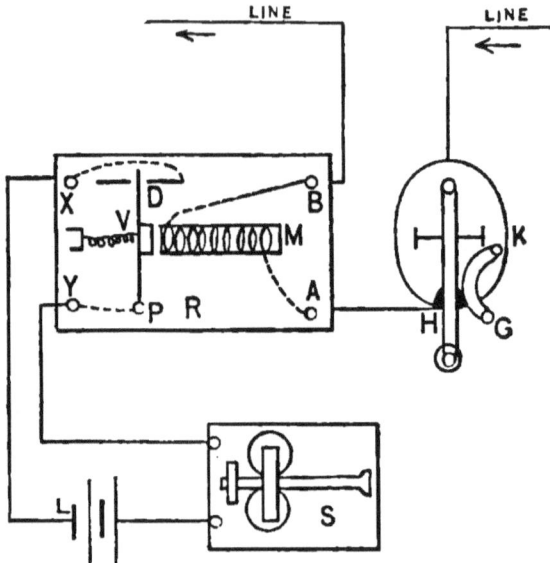

Fig. 34.

in the position shown touching the head H of the screw B (Fig. 33), and closing the circuit. From the key it passes to the relay R, entering at the binding post A, passing around the electromagnet M, and issuing at B, passing into the line to the next station. This current is furnished either by a powerful battery at one end of the line, or by a battery at each end, acting in the same direction. In front of the electromagnet M is a vibrating lever of iron or one furnished with an iron armature, pivoting at the point P. When the current passes, this lever touches the stop D and closes the local circuit $DXLSYP$ through the "sounder" S.   When the line current ceases, the lever V is drawn back by the spring and contact at D is broken. Whenever, therefore, an operator at any station opens his switch and signals, every relay on the line works, and each relay works a "sounder" through the intervention of its local battery.   As the current always runs in the same

Fig. 35.—RELAY.

direction, the relay works for every signal from which ever way it may come.   In the open-circuit system a relay is necessary for messages in each direction.   The Western Union relay is shown in Fig. 35.

The printing receiver or embosser is but little used in America, messages being read by sound.   The "sounder"

Fig. 36.—SOUNDER.

consists of two electromagnets which attract an armature

whenever the local current passes. The armature is attached to a lever, which makes a sharp click by striking against a stop whenever the armature moves. After a little practice the operator can read the message easily.

84. Faults (§ 427).

Formulas may be easily worked out for determining the position of a fault, on the supposition that the resistance of the fault is itself constant. In practice, this is seldom the case, and never so in submarine cables, as the current escaping at the fault causes electrolysis of the sea water, either depositing chloride of copper over the fault, or clearing away such deposit according as the current is positive or negative. The exact determination of the position of faults requires, therefore, great skill in making the tests and good judgment in interpreting the results obtained.

85. Simultaneouss Transmission (§ 428).

This method generally requires the use of a polarized relay. That of Siemens is probably the most easily understood. The simple form of the relay is shown in Fig. 37. S is the south pole of a steel magnet bent at a right angle. The lever aD is of soft iron pivoted at D and is of south polarity. Attached to the north pole of the steel magnet are two soft iron cores n and n', around which is coiled wire in the same circuit but in opposite directions. When no current passes the cores are of north polarity, and the oscillating lever aD is attracted to the one nearest it. If a current is sent through the coils, the cores become electromagnets of opposite polarity, and the lever then moves towards the north pole. If the current is reversed the lever moves in the opposite direction, and if the circuit is broken the lever moves towards the nearest core, as both then become north poles from the inductive action of the steel magnet. The motion of the lever is con-

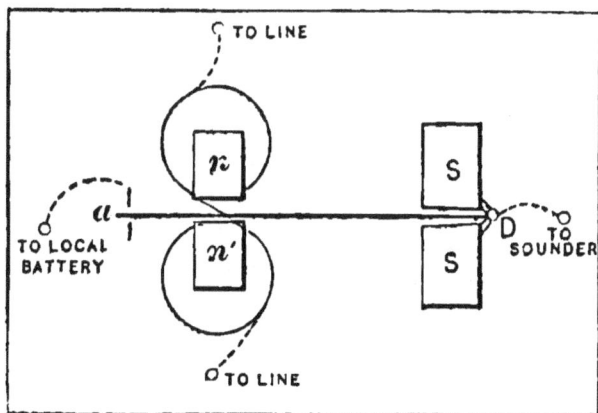

Fig. 37.

trolled by the two studs at *a*, the upper of which is connected with a local battery in circuit with a sounder, as in the common relay. The position of these studs is so regulated that the lever, even when touching the upper one and closing the local circuit, is nearer *n'* than *n*. The moment, therefore, that the line current ceases, the lever is attracted by two north poles, but moves towards the nearest, breaking the local circuit at the stud. No springs are necessary, nor does the relay require any adjustment for strength of current. From the foregoing it is seen that when the direction of the current is such as to make *n* a north pole and *n'* a south, the lever moves and closes the local circuit. When no current passes in the line, or when it passes in the opposite direction, the local circuit is open.

In sending two messages at the same time in the same direction, two keys are used, one reversing the current, sending positive or negative currents, the other sending weak or powerful. The strength of the current is, therefore, controlled by one key, its sign by the other. The method used by Edison for transmitting is shown in the figure. In the position shown the battery B has its terminals at N and P, the current passing from B through K_2 to the spring S and thence to P. If the key K' is worked currents of either polarity may be sent into the line, and passing through a polarized relay at the receiving station, a sounder in the local circuit is worked whenever a current in a given direction is transmitted by K'. The strength of the current is immaterial, the polarized relay answering only to currents in one direction. As shown in the figure, the circuit of the battery B', which is much larger than B, is open. If, however, the key K_2 is depressed the spring S comes in contact with the point m and breaks contact with n, and as it is separated from K_2 by the insulating material I, the current of B now has to pass through B' m and S to P, and is, of course, reinforced by the the powerful current of B' in the same direction. Whenever K_2 is depressed, therefore, the points N and P retain their polarity, but the current is of three or four times its original strength. In practice all contacts are made by springs, so that the circuit is never broken at K_2, but one current is followed directly by the other. The message transmitted by K_2 is received by an ordinary relay in the same circuit with the polarized relay at the receiving station, the lever of which is controlled by a spring so adjusted that the weak current of B will not cause sufficient magnetism in the elec-

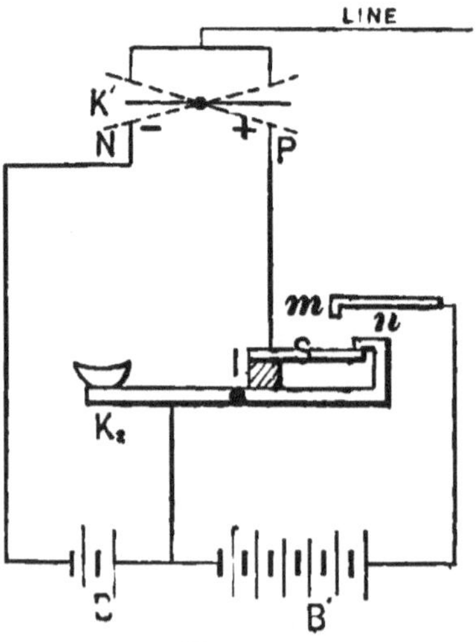

Fig. 38.

tromagnets to attract the armature against the action of the spring, but when K_2 is worked the current due to $B + B'$ easily overcomes it, whether the current be positive or negative, and the relay, therefore, transmits all signals made by K_2.

The quadruplex is merely an extension of the duplex, using the diplex or double transmission. If in Fig. 163 (Thompson) the transmitting apparatus just described is used instead of the keys R and R', and if between A and B two relays are placed in series, one an ordinary relay and the other a polarized, the figure would represent the general arrangement of Edison's quadruplex system widely used in the United States.

86. Blake's Transmitter.

In most telephone circuits, the receiving instrument is a Bell telephone, but the transmitting is a modification of Edison's telephone, known as Blake's Transmitter. The waves of sound impinge on a metallic diaphragm, causing it to press with more or less force on a carbon button. A current from a battery passes through the button and the varying pressure of the diaphragm causes a varying resistance in the circuit, and produces in the current fluctuations, corresponding in number and time to the waves of sound. If this current is passed through a Bell telephone, the message could be heard. As now used, however, the battery circuit is entirely local. In this local circuit is the primary coil of an induction coil, the secondary being in circuit with the line to the next station. Every fluctuation, therefore, in the strength of the local circuit, due to the change of pressure on the carbon button of the transmitter, induces a current in the secondary coil which works a Bell telephone at the distant station. The induction coil is small, but it causes the electromotive force of the line circuit to be much greater than that due to the battery and extends the use of the telephone to greater distances.

87. Telephone Exchanges.

The use of the telephone has been greatly extended by the system of exchanges. A large number of persons have telephone circuits to a central office, where any two circuits may be joined, thus enabling any two to converse. A great difficulty in all telephone circuits is due to induction. The instrument is so extremely delicate that any inconstant current near it induces sufficiently powerful currents in the telephone circuit to frequently obliterate a message entirely. Telegrams may be read in telephones if the telegraph and telephone circuits approach each other very closely, and telephone messages may also be heard in other circuits than that in which they are transmitted. Most of the disturbances commonly attributed to induction are, however, in all probability due to grounded telegraph circuits.

REFERENCES TO PROF. THOMPSON'S ELEMENTARY LESSONS.

§ 191, Note 26.	§ 353, Note 45, 46, 47.
§ 192, " 34.	§ 357, " 48.
§ 199, " 1.	§ 358, " 49, 50.
§ 200, " 1.	§ 360, " 51.
§ 201, " 2.	§ 361, " 52.
§ 202, " 3.	§ 362, " 53.
§ 203, " 4.	§ 364, " 54.
§ 204, " 5.	§ 367, " 55.
§ 237, " 6, 7, 8, 9, 10.	§ 371, " 56.
§ 238, " 11.	§ 372, " 57.
§ 239, " 12.	§ 374, " 58.
§ 240, " 13.	§ 375, " 76.
§ 241, " 14.	§ 376, " 77.
§ 245, " 15.	§ 377, " 78, 79.
§ 246, " 16.	§ 378, " 55.
§ 247, " 17.	§ 380, " 80.
§ 252, " 18.	§ 391,
§ 258, " 19.	§ 392, } " 59.
§ 261, " 20, 21.	§ 393,
§ 262, " 51, c.	§ 394, " 59, 60.
§ 310, " 22, 23, 24, 25, 26, 27.	§ 395, " 61.
	§ 396, " 62.
§ 311, " 28.	§ 397, " 63.
§ 312, " 23.	§ 398, " 66.
§ 313, " 29.	§ 404, " 64.
§ 314, " 30.	§ 405, " 65.
§ 315, " 31, 32.	§ 407, } " 67, 68, 69, 70.
§ 316, " 33.	§ 408,
§ 317, " 33.	§ 409, " 71.
§ 318, " 34, 35, 36, 37, 38.	§ 410, " 72.
§ 319, " 36.	§ 411, " 73, 74, 75.
§ 320, " 39.	§ 415, " 81.
§ 324, " 40.	§ 425, " 82.
§ 325a, " 41.	§ 426, " 83.
§ 327, " 42.	§ 427, " 84.
§ 338, " 42.	§ 428, " 85.
§ 351, " 43.	§ 436, " 86, 87.
§ 352, " 44.